UNCONDITIONAL QUALITY

A Harvard Business Review Paperback

Contents

Case Studies in Quality

Service Quality

The American Problem

Quality on the line

David A. Garvin

Hard new evidence on American product quality underscores the task ahead for management

Analyses of what has gone wrong with American industry have returned, time and again, to the poor quality of American-made products and to the management philosophy responsible for that quality. To date, most of the available evidence has been largely impressionistic, and few managers have felt the need to question familiar, long-established approaches to the work of manufacturing. We no longer have that excuse.

Mr. Garvin has spent several years studying the quality performance of virtually every competitor, American and Japanese, and every plant in a single but broadly representative manufacturing industry: room air conditioners. His findings document a bitter but inescapable truth. The competitively significant variation in levels of quality performance is immense. But the truth is also heartening: superior levels of performance come not from national traits or cultural advantages but from sound management practices deliberately and systematically applied. If nothing else, the data reported here should force American executives to rethink their approaches to product quality.

Mr. Garvin is assistant professor of business administration at the Harvard Business School. A previous HBR article, "Managing As If Tomorrow Mattered" (May-June 1982), which he wrote with Robert Hayes, won the McKinsey Award.

Photograph by Keith W. Jenkins.

When it comes to product quality, American managers still think the competitive problem much less serious than it really is. Because defining the term accurately within a company is so difficult (is quality a measure of performance, for example, or reliability or durability), managers often claim they cannot know how their product quality stacks up against that of their competitors, who may well have chosen an entirely different quality "mix." And since any comparisons are likely to wind up as comparisons of apples with oranges, even a troubling variation in results may reflect only a legitimate variation in strategy. Is there, then, a competitive problem worth worrying about?

I have recently completed a multiyear study of production operations at all but one of the manufacturers of room air conditioners in both the United States and Japan (details of the study are given in the insert, *Research methods*). Each manufacturer uses a simple assembly-line process; each uses much the same manufacturing equipment; each makes an essentially standardized product. No apples versus oranges here: the comparison is firmly grounded. And although my data come from a single industry, both that industry's manufacturing process and its managers' range of approaches to product quality give these findings a more general applicability.

The shocking news, for which nothing had prepared me, is that the failure rates of products from the highest-quality producers were between 500

Author's note: I thank Professor Robert Stobaugh for his helpful comments on an earlier draft, the Division of Research at the Harvard Business School for financial support, and the Nomura School of Advanced Management in Tokyo for arranging and coordinating my trip to Japan.

and *1,000* times less than those of products from the lowest. The "between 500 and 1,000" is not a typographical error but an inescapable fact. There is indeed a competitive problem worth worrying about.

Measuring quality

Exhibit I presents a composite picture of the quality performance of U.S. and Japanese manufacturers of room air conditioners. I have measured quality in two ways: by the incidence of "internal" and of "external" failures. Internal failures include all defects observed (either during fabrication or along the assembly line) before the product leaves the factory; external failures include all problems incurred in the field after the unit has been installed. As a proxy for the latter, I have used the number of service calls recorded during the product's first year of warranty coverage because that was the only period for which U.S. and Japanese manufacturers offered comparable warranties.

Measured by either criterion, Japanese companies were far superior to their U.S. counterparts: their average assembly-line defect rate was almost 70 times lower and their average first-year service call rate nearly 17 times better. Nor can this variation in performance be attributed simply to differences in the number of minor, appearance-related defects. Classifying failures by major functional problems (leaks, electrical) or by component failure rates (compressors, thermostats, fan motors) does not change the results.

More startling, on both internal and external measures, the poorest Japanese company typically had a failure rate less than half that of the best U.S. manufacturer. Even among the U.S. companies themselves, there was considerable variation. Assembly-line defects ranged from 7 to 165 per 100 units—a factor of 20 separating the best performer from the worst—and service call rates varied by a factor of 5.

For ease of analysis, I have grouped the companies studied according to their quality performance (see the *Appendix*). These groupings illustrate an important point: quality pays. *Exhibit II*, for example, presents information on assembly-line productivity for each of these categories and shows that the highest-quality producers were also those with the highest output per man-hour. On the basis of the number of direct labor hours actually worked on the assembly line, productivity at the best U.S. companies was five times higher than at the worst.

Measuring productivity by "standard output" (see *Exhibit II*) blurs the picture somewhat. Although the Japanese plants maintain a slight edge over the best U.S. plants, categories of performance tend to overlap. The figures based on standard output, however, are rather imperfect indicators of productivity—for example, they fail to include overtime or rework hours and so overstate productivity levels, particularly at the poorer companies, which devote more of their time to correcting defects. Thus, these figures have less significance than do those based on the number of hours actually worked.

Note carefully that the strong association between productivity and quality is not explained by differences in technology or capital intensity, for most of the plants employed similar manufacturing techniques. This was especially true of final assembly, where manual operations, such as hand brazing and the insertion of color-coded wires, were the norm. Japanese plants did use some automated transfer lines and packaging equipment, but only in compressor manufacturing and case welding was the difference in automation significant.

The association between cost and quality is equally strong. Reducing field failures means lower warranty costs, and reducing factory defects cuts expenditures on rework and scrap. As *Exhibit III* shows, the Japanese manufacturers incurred warranty costs averaging 0.6% of sales; at the best American companies, the figure was 1.8%; at the worst 5.2%.

In theory, low warranty costs might be offset by high expenditures on defect prevention: a company could spend enough on product pretesting or on inspecting assembled units before shipment to wipe out any gains from improved warranty costs. Figures on the total costs of quality, however, which include expenditures on prevention and inspection as well as the usual failure costs of rework, scrap, and warranties, lead to the opposite conclusion. In fact, the total costs of quality incurred by Japanese producers were less than one-half the failure costs incurred by the best U.S. companies.

The reason is clear: failures are much more expensive to fix after a unit has been assembled than before. The cost of the extra hours spent pretesting a design is cheap compared with the cost of a product recall; similarly, field service costs are much higher than those of incoming inspection. Among manufacturers of room air conditioners, the Japanese—even with their strong commitment to design review, vendor selection and management, and in-process inspection—still have the lowest overall quality costs.

Nor are the opportunities for reduction in quality costs confined to this industry alone. A recent survey[1] of U.S. companies in ten manufacturing sectors found that total quality costs averaged 5.8% of sales—for a $1 billion corporation, some $58 million

---

1 "Quality Cost Survey,"
Quality,
June 1977, p. 20.

Exhibit I

Quality in the room air conditioning industry
1981-1982*

	Internal failures				External failures			
	Fabrication: coil leaks per 100 units	Assembly-line defects per 100 units			Service call rate per 100 units under first-year warranty coverage			
		Total	Leaks	Electrical	Total	Compressors	Thermostats	Fan motors
Median								
United States	4.4	63.5	3.1	3.3	10.5†	1.0	1.4	.5
Japan	< .1	.95	.12	.12	.6	.05	.002	.028
Range								
United States	.1-9.0	7-165	1.3-34	.9-34	5.3-26.5	.5-3.4	.4-3.6	.2-2.6
Japan	.03-.4	.15-3.0	.0015-.5	.0015-1.0	.04-2.0	.002-.1	0-.03	.001-.2

*Although most companies reported total failure rates for 1981 or 1982, complete data on component failure rates were often available only for earlier years. For some U.S. companies, 1979 or 1980 figures were employed. Because there was little change in U.S. failure rates during this period, the mixing of data from different years should have little effect.

†Service call rates in the United States normally include calls where no product problems were found ("customer instruction" calls); those in Japan do not. I have adjusted the U.S. median to exclude these calls; without the adjustments, the median U.S. service call rate would be 11.4 per 100 units. Figures for the range should be adjusted similarly, although the necessary data were not available from the U.S. companies with the highest and lowest service call rates.

Exhibit II

Quality and productivity

Grouping of companies by quality performance	Units produced per assembly-line direct labor man-hour actual hours*		Units produced per assembly-line direct labor man-hour standard output†	
	Median	Range	Median	Range
Japanese manufacturers	NA‡	NA	1.8	1.4-3.1
Best U.S. plants	1.7	1.7§	1.7	1.4-1.9
Better U.S. plants	.9	.7-1.0	1.1	.8-1.2
Average U.S. plants	1.0	.6-1.2	1.1	1.1-1.7
Poorest U.S. plants	.35	.35§	1.3	.8-1.6

*Direct labor hours have been adjusted to include only those workers involved in assembly (i.e., where inspectors and repairmen were classified as direct labor, they have been excluded from the totals).

†Computed by using the average cycle time to derive a figure for hourly output, and then dividing by the number of assembly-line direct laborers (excluding inspectors and repairmen) to determine output per man-hour.

‡NA = not available.

§In this quality grouping, man-hour data were only available from a single company.

per year primarily in scrap, rework, and warranty expenses. Shaving even a tenth of a percentage point off these costs would result in an annual saving of $1 million.

Other studies, which use the PIMS data base, have demonstrated a further connection among quality, market share, and return on investment.[2] Not only does good quality yield a higher ROI for any given market share (among businesses with less than 12% of the market, those with inferior product quality averaged an ROI of 4.5%, those with average product quality an ROI of 10.4%, and those with superior product quality an ROI of 17.4%); it also leads directly to market share gains. Those businesses in the PIMS study that improved in quality during the 1970s increased their market share five to six times faster than those that declined—and three times faster than those whose quality remained unchanged.

The conclusion is inescapable: improving product quality is a profitable activity. For managers, therefore, the central question must be: What makes for successful quality management?

Sources of quality

Evidence from the room air conditioning industry points directly to the practices that the quality leaders, both Japanese and American, have employed. Each of these areas of effort—quality programs, policies, and attitudes; information systems; product design; production and work force policies; and vendor management—has helped in some way to reduce defects and lower field failures.

Programs, policies & attitudes

The importance a company attaches to product quality often shows up in the standing of its quality department. At the poorest performing plants in the industry, the quality control (QC) manager invariably reported to the director of manufacturing or engineering. Access to top management came, if at all, through these go-betweens, who often had very different priorities from those of the QC manager. At the best U.S. companies, especially those with low service call rates, the quality department had more visibility.

Several companies had vice presidents of quality; at the factory level each head of QC reported directly to the plant manager. Japanese QC managers also reported directly to their plant managers.

Of course, reporting relationships alone do not explain the observed differences in quality performance. They do indicate, however, the seriousness that management attaches to quality problems. It's one thing to say you believe in defect-free products, but quite another to take time from a busy schedule to act on that belief and stay informed. At the U.S. company with the lowest service call rate, the president met weekly with all corporate vice presidents to review the latest service call statistics. Nobody at that company needed to ask whether quality was a priority of upper management.

How often these meetings occurred was as important as their cast of characters. Mistakes do not fix themselves; they have to be identified, diagnosed, and then resolved through corrective action. The greater the frequency of meetings to review quality performance, the fewer undetected errors. The U.S. plants with the lowest assembly-line defect rates averaged ten such meetings per month; at all other U.S. plants, the average was four. The Japanese companies reviewed defect rates daily.

Meetings and corrective action programs will succeed, however, only if they are backed by genuine top-level commitment. In Japan, this commitment was deeply ingrained and clearly communicated. At four of the six companies surveyed, first-line supervisors believed product quality—not producing at low cost, meeting the production schedule, or improving worker productivity—was management's top manufacturing priority. At the other two, quality ranked a close second.

The depth of this commitment became evident in the Japanese practice of creating internal consumer review boards. Each of the Japanese producers had assembled a group of employees whose primary function was to act as typical consumers and test and evaluate products. Sometimes the products came randomly from the day's production; more frequently, they represented new designs. In each case, the group had final authority over product release. The message here was unmistakable: the customer—not the design staff, the marketing team, or the production group— had to be satisfied with a product's quality before it was considered acceptable for shipment.

By contrast, U.S. companies showed a much weaker commitment to product quality. At 9 of the 11 U.S. plants, first-line supervisors told me that their managers attached far more importance to meeting the production schedule than to any other manufacturing objective. Even the best performers showed no consistent relationship between failure rates and supervisors' perceptions of manufacturing priorities.

2 Sidney Schoeffler, Robert D. Buzzell, and Donald F Heany, "Impact of Strategic Planning on Profit Performance," HBR March-April 1974, p. 137; and

Robert D. Buzzell and Frederik D. Wiersema, "Successful Share-Building Strategies," HBR January-February 1981, p. 135.

Exhibit III   **Quality and costs**

Grouping of companies by quality performance	Warranty costs as a percentage of sales*		Total cost of quality (Japanese companies) and total failure costs (U.S. companies) as a percentage of sales†	
	Median	Range	Median	Range
Japanese manufacturers	.6%	.2-1.0%	1.3%	.7-2.0%
Best U.S. plants	1.8	1.1-2.4	2.8	2.7-2.8
Better U.S. plants	2.4	1.7-3.1	3.4	3.3-3.5
Average U.S. plants	2.2	1.7-4.3	3.9	2.3-5.6
Poorest U.S. plants	5.2%	3.3-7.0%	>5.8%	4.4->7.2%

*Because most Japanese air conditioners are covered by a three-year warranty while most U.S. units are covered by a warranty of five years, these figures somewhat overstate the Japanese advantage. The bias is unlikely to be serious, however, because second- to fifth-year coverage in the United States and second- to third-year coverage in Japan are much less inclusive – and therefore, less expensive – than first-year coverage. For example, at U.S. companies second- to fifth-year warranty costs average less than one-fifth of first-year expenses.

†Total cost of quality is the sum of all quality-related expenditures, including the costs of prevention, inspection, rework, scrap, and warranties. The Japanese figures include expenditures in all of these categories, while the U.S. figures, because of limited data, include only the costs of rework, scrap, and warranties (failure costs). As a result, these figures understate total U.S. quality costs relative to those of the Japanese.

Exhibit IV   **Quality and product stability**

Grouping of companies by quality performance	Median number of design changes per year	Median number of models	Median number of design changes per model*	Median percentage that peak production exceeded low production†
Japanese manufacturers	NA‡	80	NA	170%
Best U.S. plants	43	56	.8	27
Better U.S. plants	150	81	1.9	63
Average U.S. plants	400	126	3.2	50
Poorest U.S. plants	133	41	3.2	100%

*Column 1 divided by column 2.

†The figures in this column were derived by dividing each plant's largest daily output for the year by its smallest (nonzero) output for the year.

‡NA = not available.

What commitment there was stemmed from the inclusion (or absence) of quality in systems of performance appraisal. Two of the three companies with the highest rates of assembly-line defects paid their workers on the basis of total output, not defect-free output. Is it any wonder these employees viewed defects as being of little consequence? Not surprisingly, domestic producers with low failure rates evaluated both supervisors and managers on the quality of output—supervisors, in terms of defect rates, scrap rates, and the amount of rework attributable to their operations; managers, in terms of service call rates and their plants' total costs of quality.

These distinctions make good sense. First-line supervisors play a pivotal role in managing the production process, which is responsible for internal failures, but have little control over product design, the quality of incoming materials, or other factors that affect field performance. These require the attention of higher level managers, who can legitimately be held responsible for coordinating the activities of design, purchasing, and other functional groups in pursuit of fewer service calls or reduced warranty expenses.

To obtain consistent improvement, a formal system of goal setting is necessary.[3] Only three U.S. plants set annual targets for reducing field failures. Between 1978 and 1981, these three were the only ones to cut their service call rates by more than 25%; most of the other U.S. plants showed little or no change. All the Japanese companies, however, consistently improved their quality—in several cases, by as much as 50%—and all had elaborate companywide systems of goal setting.

From the corporate level at these companies came vague policy pronouncements ("this year, let the customer determine our quality"), which were further defined by division heads ("reduced service call rates are necessary if we are to lower costs") and by middle managers ("compressor failures are an especially serious problem that must be addressed"). Actual quantitative goals ("improve compressor reliability by 10%") were often set by foremen or workers operating through quality control circles. The collaborative nature of this goal-setting process helped these targets earn wide support.

At the final—or first—level of goal setting, specificity matters. Establishing an overall target for an assembly-line defect rate without specifying more detailed goals by problem category, such as leaks or electrical problems, is unlikely to produce much improvement. A number of U.S. plants have tried this approach and failed. Domestic producers with the lowest defect rates set their overall goals last. Each inspection point along the assembly line had a target of its own, which was agreed on by representatives of the quality and manufacturing departments. The sum of these individual targets established the overall goal for the assembly line. As a result, responsibility for quality became easier to assign and progress easier to monitor.

Information systems

Successful monitoring of quality assumes that the necessary data are available, which is not always true. Without specific and timely information on defects and field failures, improvements in quality are seldom possible. Not surprisingly, at the poorest U.S. companies information on defects and field failures was virtually nonexistent. Assembly-line defects and service call rates were seldom reported. "Epidemic" failures (problems that a large proportion of all units had in common) were widespread. Design flaws remained undetected. At one domestic producer, nearly a quarter of all 1979-1981 warranty expenses came from problems with a single type of compressor.

Other companies reported more extensive quality information—daily and weekly defect rates as well as quarterly and, occasionally, monthly service call rates. These variations in the level of reporting detail correlated closely with differences in quality performance. Among average U.S. performers, for example, quality reports were quite general. Data on assembly line defects gave no breakdowns by inspection point; data on field failures were for entire product lines, not for separate models. Without further refinement, such data cannot isolate quality problems.

A 10% failure rate for a product line can mean a number of things: that all models in the line fail to perform 10% of the time, with no single problem standing out; that several models have a 5% failure rate and one a 30% rate, which suggests a single problem of epidemic proportions; or anything in between. There is no way of distinguishing these cases on the basis of aggregate data alone. What is true of goal setting is equally true of reporting systems: success requires mastering the details.

The best U.S. companies reported defect rates for each inspection point on the assembly line and field failure rates by individual model. The Japanese not only collected information that their U.S. counterparts ignored, such as failure rates in the later years of a product's life; they also insisted on extreme precision in reporting. At one company, repairmen had to submit reports on every defective unit they fixed. In general, it was not unusual for Japanese managers to be able to identify the 30 different ways in which Switch X had failed on Model Y during the last several years. Nor did they have to wait for such information.

3 For a summary of evidence on this point, see Edwin A. Locke et al., "Goal Setting and Task Performance: 1969-1980," *Psychological Bulletin,* vol. 90, no. 1, p. 125.

Research methods

This article is based mainly on data collected in 1981 and 1982 from U.S. and Japanese manufacturers of room air conditioners. I selected that industry for study for a number of reasons: it contains companies of varying size and character, which implies a wide range of quality policies and performance; its products are standardized, which facilitates inter-company comparisons; and it employs a simple assembly-line process, which is representative of many other mass production industries.

Nine of the ten U.S. companies in the industry and all seven of the Japanese companies participated in the study. They range in size from small air conditioning specialists with total sales of under $50 million to large home appliance manufacturers with annual sales of more than $200 million in this product line alone. Taken together, they account for approximately 90% of U.S. industry shipments and 90% of Japanese industry shipments. I have collected data separately for each plant (two of the American companies operate two plants apiece; otherwise, each company employs only a single plant). Of the 18 plants studied, 11 are American and 7 Japanese.

Once U.S. companies had agreed to participate in the study, I sent them a questionnaire requesting background information on their product line, production practices, vendor management practices, quality policies, and quality performance. I then visited them all in order to review the questionnaire results, collect additional data, tour the factories, and conduct interviews with key personnel. The interviews were open-ended and unstructured, although I posed similar questions at each company. A typical visit included interviews with managers in the quality, manufacturing, purchasing, engineering, and service departments, as well as several hours spent walking the production floor.

Preliminary analysis of the interviews and questionnaires showed that companies neither employed the same conventions in reporting data nor answered questions in the same degree of detail. I therefore sent each company its own set of follow-up questions to fill in these gaps and to make the data more comparable across companies. In addition, I requested each company to administer a brief questionnaire on quality attitudes to each of its first-line production supervisors.

I followed a similar approach with the Japanese manufacturers, although time constraints limited the amount of information that I could collect. All questionnaires were first translated into Japanese and mailed to the participating companies. Six of the seven companies completed the same basic quality questionnaire as did their American counterparts; the same companies also administered the survey on quality attitudes to a small number of their first-line supervisors. With the aid of a translator, I conducted on-site interviews at all the companies and toured six of the plants.

Service call statistics in the United States took anywhere from one month to one year to make the trip from the field to the factory; in Japan, the elapsed time averaged between one week and one month. Differences in attitude are part of the explanation. As the director of quality at one Japanese company observed, field information reached his company's U.S. subsidiaries much more slowly than it did operations in Japan—even though both employed the same system for collecting and reporting data.

Product design

Room air conditioners are relatively standardized products. Although basic designs in the United States have changed little in recent years, pressures to improve energy efficiency and to reduce costs have resulted in a stream of minor changes. On the whole, these changes have followed a common pattern: the initiative came from marketing; engineering determined the actual changes to be made and then pretested the new design; quality control, manufacturing, purchasing, and other affected departments signed off; and, where necessary, prototypes and pilot production units were built.

What did differ among companies was the degree of design and production stability. As *Exhibit IV* indicates, the U.S. plants with the lowest failure rates made far fewer design changes than did their competitors.

Exhibit IV conveys an important message. Variety, at least in America, is often the enemy of quality. Product proliferation and constant design change may keep the marketing department happy, but failure rates tend to rise as well. By contrast, a limited product line ensures that workers are more familiar with each model and less likely to make mistakes. Reducing the number of design changes allows workers to devote more attention to each one. Keeping production level means less reliance on a second shift staffed by inexperienced employees.

The Japanese, however, have achieved low failure rates even with relatively broad product lines and rapidly changing designs. In the room air conditioning industry, new designs account for nearly a third of all models offered each year, far more than in the United States. The secret: an emphasis on reliability engineering and on careful shakedowns of new designs before they are released.

Reliability engineering is nothing new; it has been practiced by the aerospace industry in this country for at least 20 years. In practice, it involves building up designs from their basic components, determining the failure probabilities of individual systems and subsystems, and then trying to strengthen the weak links in the chain by product redesign or by incorporating more reliable parts. Much of the effort is focused up front, when a product is still in blueprint or prototype form. Managers use statistical techniques to predict reliability over the product's life and subject preliminary designs to exhaustive stress and life testing to collect information on potential failure modes. These data form the basis for continual product improvement.

Only one American maker of room air conditioners practiced reliability engineering, and its failure rates were among the lowest observed. All of the Japanese companies, however, placed considerable emphasis on these techniques. Their designers were, for example, under tremendous pressure to reduce the number of parts per unit; for a basic principle of reliability engineering is that, everything else being equal, the fewer the parts, the lower the failure rate.

Japanese companies worked just as hard to increase reliability through changes in components. They were aided by the Industrial Engineering Bureau of Japan's Ministry of International Trade and Industry (MITI), which required that all electric and electronic components sold in the country be tested for reliability and have their ratings on file at the bureau. Because this information was publicly available, designers no longer needed to test components themselves in order to establish reliability ratings.

An emphasis on reliability engineering is also closely tied to a more thorough review of new designs before units reach production. American manufacturers usually built and tested a single prototype before moving to pilot production; the Japanese often repeated the process three or four times.

Moreover, all affected departments—quality control, purchasing, manufacturing, service, and design engineering—played an active role at each stage of the review process. American practice gave over the early stages of the design process almost entirely to engineering. By the time other groups got their say, the process had gained momentum, schedules had been established, and changes had become difficult to make. As a result, many a product that performed well in the laboratory created grave problems on the assembly line or in the field.

Production & work force policies

The key to defect-free production is a manufacturing process that is "in control"—machinery and equipment well maintained, workplaces clean and orderly, workers well trained, and inspection procedures suited to the rapid detection of deviations. In each of these areas, the Japanese were noticeably ahead of their American competitors.

Training of the labor force, for example, was extensive, even for employees engaged in simple jobs. At most of the Japanese companies, preparing new assembly-line workers took approximately six months, for they were first trained for all jobs on the line. American workers received far less instruction (from several hours to several days) and were usually trained for a single task. Not surprisingly, Japanese workers were much more adept at tracking down quality problems originating at other work stations and far better equipped to propose remedial action.

Instruction in statistical quality control techniques supplemented the other training offered Japanese workers. Every Japanese plant relied heavily on these techniques for controlling its production process. Process control charts, showing the acceptable quality standards of various fabrication and assembly-line operations, were everywhere in general use. Only one U.S. plant—the one with the lowest defect rate—had made a comparable effort to determine the capabilities of its production process and to chart the results.

Still, deviations will occur, and thorough and timely inspection is necessary to ferret them out quickly. Japanese companies therefore employed an inspector for every 7.1 assembly-line workers (in the United States the ratio was 1:9.5). The primary role of these inspectors was to monitor the production process for stability; they were less "gatekeepers," weeding out defective units before shipment, than providers of information. Their tasks were considered especially important where manual operations were common and where inspection required sophisticated testing of a unit's operating characteristics.

On balance, then, the Japanese advantage in production came less from revolutionary technology than from close attention to basic skills and to the reduction of all unwanted variations in the manufacturing process. In practice, this approach can produce dramatic results. Although new model introductions and assembly-line changeovers at American companies boosted defect rates, at least until workers became familiar with their new assignments, Japanese companies experienced no such problems.

Before every new model introduction, Japanese assembly-line workers were thoroughly trained in their new tasks. After-hours seminars explained the product to the work force, and trial runs were common. During changeovers, managers kept workers informed of the models slated for production each day, either through announcements at early morning meetings or by sending assembled versions of the new model down the line 30 minutes before the change was to take place, together with a big sign saying "this model comes next." American workers generally received much less information about changeovers. At the plant with the highest defect rate in the industry, communication about changeovers was limited to a single small chalkboard, listing the models to be produced each day, placed at one end of the assembly line.

The Japanese system of permanent employment also helped to improve quality. Before they are fully trained, new workers often commit unintentional errors. Several American companies observed that their workers' inexperience and lack of familiarity with the product line contributed to their high defect rates. The Japanese, with low absenteeism and turnover, faced fewer problems of this sort. Japanese plants had a median turnover of 3.1%; the comparable figure for U.S. plants was two times higher. Even more startling were the figures on absenteeism: a median of 3.1% for American companies and *zero* for the Japanese.

In addition, because several of the U.S. plants were part of larger manufacturing complexes linked by a single union, they suffered greatly from "bumping." A layoff in one part of the complex would result in multiple job changes as workers shifted among plants to protect seniority rights. Employees whose previous experience was with another product would suddenly find themselves assembling room air conditioners. Sharp increases in defects were the inevitable result.

Vendor management

Without acceptable components and materials, no manufacturer can produce high-quality products. As computer experts have long recognized, "garbage in" means "garbage out." Careful selection and monitoring of vendors is therefore a necessary first step toward ensuring reliable and defect-free production.

At the better U.S. companies, the quality department played a major role in vendor selection by tempering the views of the engineering ("do their samples meet our technical specifications") and purchasing ("is that the best we can do on price") departments. At the very best companies, however, purchasing departments independently ranked quality as their primary objective. Buyers received instruction in the concepts of quality control; at least one person had special responsibility for vendor quality management; goals were set for the quality of incoming components and materials; and vendors' shipments were carefully monitored.

Purchasing departments at the worst U.S. companies viewed their mission more narrowly: to obtain the lowest possible price for technically acceptable components. Site visits to new vendors were rarely made, and members of the purchasing department seldom got involved in the design review process. Because incoming inspection was grossly understaffed (at one plant, two workers were responsible for reviewing 14,000 incoming shipments per year), production pressures often allowed entire lots to come to the assembly line uninspected. Identification of defective components came, if at all, only after they had been incorporated into completed units. Inevitably, scrap and rework costs soared.

In several Japanese companies incoming materials arrived directly at the assembly line without inspection. New vendors, however, first had to pass rigorous tests: their products initially received 100% inspection. Once all problems were corrected, sampling inspection became the norm. Only after an extended period without a rejection were vendors allowed to send their products directly to the assembly line. At the first sign of deterioration in vendor performance, more intensive inspection resumed.

In this environment, inspection was less an end in itself than a means to an end. Receiving inspectors acted less as policemen than as quality consultants to the vendor. Site visits, for example, were mandatory when managers were assessing potential suppliers and continued for as long as the companies did business together. Even more revealing, the selection of vendors depended as much on management philosophy, manufacturing capability, and depth of commitment to quality as on price and delivery performance.

Closing the gap

What, then, is to be done? Are American companies hopelessly behind in the battle for superior quality? Or is an effective counterattack possible?

Although the evidence is still fragmentary, there are a number of encouraging signs. In 1980, when Hewlett-Packard tested 300,000 semiconductors from three U.S. and three Japanese suppliers, the Japanese chips had a failure rate one-sixth that of the American chips. When the test was repeated two years later, the U.S. companies had virtually closed the gap. Similar progress is evident in automobiles. Ford's Ranger trucks, built in Louisville, Kentucky, offer an especially dramatic example. In just three years, the number of "concerns" registered by the Louisville plant (the automaker's measure of quality deficiencies as recorded at monthly audits) dropped to less than one-third its previous high. Today, the Ranger's quality is nearly equal that of Toyota's SR5, its chief Japanese rival.

But in these industries, as with room air conditioners, quality improvement takes time. The "quick fix" provides few lasting gains. What is needed is a long-term commitment to the fundamentals— working with vendors to improve their performance, educating and training the work force, developing an accurate and responsive quality information system, setting targets for quality improvement, and demonstrating interest and commitment at the very highest levels of management. With their companies' futures on the line, managers can do no less.

[See Appendix on following page]

Appendix

Classifying plants by quality performance

To identify patterns of behavior, I first grouped U.S. plants into categories according to their quality performance on two dimensions – internal quality (defect rates in the factory) and external quality (failure rates in the field).

Table A presents the basic data on external quality. I measured field performance in two ways: by the service call rate for units under first-year warranty coverage (the total number of service calls recorded in 1981 divided by the number of units in the field with active first-year warranties) and by the service call rate for units under first-year warranty coverage less "customer instruction calls" (only those service calls that resulted from a faulty unit, not from a customer who was using the unit improperly or had failed to install it correctly).

The second measure was necessary because companies differed in their policies toward customer instruction calls. Some reimbursed repairmen for these calls without argument; others did their best to eliminate such calls completely. An accurate assessment of product performance required the separation of these calls from problems that reflect genuinely defective units.

I classified plants on the basis of their rankings on the second of the two measures in *Table A,* and then grouped them according to their actual levels of field failures. In most cases, the dividing lines were clear, although there were some borderline cases. Plant 8, for example, had a total service call rate well above the industry median, yet after subtracting customer instruction calls, its failure rate differed little from the other average performers. Because this second figure more accurately reflects the rate of product malfunction, I treated Plant 8 as having average, rather than poor, external quality. A number of companies with high failure rates did not break out customer instruction calls. I have treated them as having poor external quality because their customer instruction calls would have to have been two or three times as frequent as the highest rate recorded in 1981 for them to have warranted an average ranking.

Table A — Field performance for U.S. plants in 1981

Plant	Service call rate, first year warranty coverage		Service call rate less "customer instruction" calls	
	Percentage	Rank	Percentage	Rank
1	5.3%	1	< 5.3%	1
2	8.7	2	< 8.7	2,3
3	9.2	3	5.6	2,3
4	10.5	4	9.8	5
5	11.1	5	9.3	4
6	11.4	6	10.5	6
7	12.6	7	10.5	6
8	16.2	8	11.8	8
9	17.5	9	13.8	9
10	22.9	10	< 22.9	10
11	26.5%	11	< 26.5%	11

Ranking of plants on field performance
external quality

	Good	Average	Poor
Plant number	1, 2, 3	4, 5, 6, 7, 8	9, 10, 11

I followed a similar procedure in classifying plants on internal quality. Because companies differed in how they defined and recorded defects (some noted every single product flaw; others were interested only in major malfunctions), I employed several indexes to ensure consistency. The results are displayed in *Table B*. I ranked companies first by their total assembly-line defect rates (every defect recorded at every station along the assembly line divided by the number of units produced) and then by the number of defects requiring off-line repair. The second index compensates for the different definitions just noted, for it more accurately reflects the incidence of serious problems. Minor adjustments and touch-ups can generally be made without pulling a unit off the line; more serious problems normally require off-line repair. Measured on this basis, the high total defect rates of Plants 1 and 9 appear to be much less of a problem.

Because several companies had to estimate the off-line repair rate, I used a third index, the number of repairmen per assembly-line direct laborer, to measure defect seriousness. The proportion of the work force engaged in repair activities, including workers assigned to separate rework lines and to rework activities in the warehouse, is likely to correlate well with the incidence of serious defects, for more serious problems usually require more time to correct and necessitate a larger repair staff. This measure provides important additional information, confirming the conclusions about Plant 1 (its high total defect rate appears to include a large number of minor problems) but contradicting those about Plant 9 (its large number of repairmen suggests that defects are, in fact, a serious problem, despite the small proportion of units requiring off-line repair).

I assigned plants to groups using much the same procedure as before. I first computed a composite ranking for each plant by averaging together the three rankings of *Table B*. Dividing lines between groups followed the absolute levels of the indexes for each plant. Once again, some judgment was involved, particularly for Plants 4, 5, and 9. Plants 5 and 9 were borderline cases, candidates for ranking as either average or poor internal quality. I classified the former as average, even though its overall rank was low, because its absolute scores on the first two measures were quite close to the median. I classified the latter as poor because its absolute scores on both the first and the third measures were so high. Plant 4 presented a different problem, for it provided no information at all on assembly-line defects. Rather than classifying the plant on the basis of the third index alone, I employed supplementary data. Based on its

Table B — Internal quality for U.S. plants in 1981

Plant	Assembly-line defects per 100 units		Assembly-line defects per 100 units requiring off-line repair		Repairmen per assembly-line direct laborer	
	Number	Rank	Number	Rank	Number	Rank
1	150	9	34	5,6	.06	3
2	7	1	7	1	.05	2
3	10	2	10	3	.04	1
4	NA*	NA	NA	NA	.09	8
5	57	5	47	7	.13	9
6	70	6	67	8	.06	3
7	26	4	7	1	.08	6
8	18	3	11	4	.08	6
9	>100	7	> 30	5,6	.16	11
10	165	10	165	10	.13	9
11	135	8	> 68	9	.07	5

*NA = not available.

Ranking of plants on internal quality

	Good	Average	Poor
Plant number	2, 3, 7, 8	1, 4(?), 5, 6	9, 10, 11

defect rate at the end-of-the-line quality audit and its rework and scrap costs as a percentage of sales, both of which were quite close to figures reported by other companies with average internal quality, Plant 4 showed up as an average performer.

Table C combines the results of the previous two tables. Overall quality rankings appear for each plant. In most cases, success on internal quality implied success on external measures, although the correlation is not perfect, as Plants 1, 7, and 8 demonstrate. The Japanese plants are in a category of their own, for on both internal and external measures they are at least twice as good as the best U.S. plant.

Table C — Classification of plants on internal and external quality

Poor U.S. plants ▨
Average U.S. plants ▨
Better U.S. plants ▨
Best U.S. plants ■

Internal quality	External quality			
	Poor	Fair	Good	Excellent
Poor	Plants 9, 10, 11			
Fair		Plants 4, 5, 6	Plant 1	
Good		Plants 7, 8	Plants 2, 3	
Excellent				All Japanese plants

Reprint 83505

Solutions for
Unconditional Quality

Harvard Business School **9-687-011**

A Note on Quality: The Views of Deming, Juran, and Crosby

Associate for Case Development **Artemis March** *prepared this note under the supervision of Professor* **David A. Garvin** *as the basis for class discussion.*

During the 1980s concerns about American competitiveness steered many U.S. companies to a new interest in quality. The three leading "quality gurus" were W. Edwards Deming, Joseph Juran, and Philip Crosby. Each was an active consultant, lecturer, and author, with years of experience. Deming and Juran were in their eighties and had been enormously influential in Japan; Crosby was in his sixties and had worked previously at ITT as vice president of quality. Each had developed his own distinctive approach to quality management.

Deming   *SPC*

W. Edwards Deming was widely credited with leading the Japanese quality revolution. The Japanese began to heed his advice on statistical process control (SPC) and problem-solving techniques in 1950, but 30 years passed before American businesses began to respond. By then, Deming's message to managers was blunt: "The basic cause of sickness in American industry and resulting unemployment is failure of top management to manage."[1]

---

1. W. Edwards Deming, *Quality, Productivity, and Competitive Position* (Cambridge, MA: Massachusetts Institute of Technology, Center for Advanced Engineering Study, 1982), p. i.

Known to dismiss client companies that did not change, he stated, "I give 'em three years. I've got to see a lot happen."[2] Best efforts were not enough; a program was needed, and it had to be adopted wholeheartedly:

> Everyone doing his best is not the answer. It is necessary that people know what to do. Drastic changes are required. The responsibility for change rests on management. The first step is to learn how to change.[3]

What Deming then expected from his clients was summarized in a 14-point program (see *Exhibit 1*).

To begin, managers had to put aside their preoccupation with today to make sure there was a tomorrow. They had to orient themselves to continuous improvement of products and services to meet customers' needs and stay ahead of the competition. They had to innovate constantly and commit resources to support innovation and continuous quality improvement. They had to build quality in. They had to break down department and worker-supervisor barriers. They had to rid themselves of numerical targets and quotas and instead had to concentrate on improving processes, giving workers clear standards for acceptable work, as well as the tools needed to achieve it. Finally, they had to create a climate free of finger pointing and fear, which block cooperative identification and solution of problems.

If management committed itself to this new order, Deming argued, productivity as well as quality would improve. Contrary to conventional wisdom in the United States, quality and productivity were not to be traded off against each other. Rather, productivity was a by-product of quality and of doing the job right the first time:

> Improvement of the process increases uniformity of product, reduces rework and mistakes, reduces waste of manpower, machine-time, and materials, and thus increases output with less effort. Other benefits of improved quality are lower costs, . . . happier people on the job, and more jobs, through better competitive position of the company.[4]

Because management was responsible, in Deming's view, for 85% of all quality problems, management had to take the lead in changing the systems and processes that created those problems. For example, consistent quality of incoming materials and components could not be expected when buyers were told to shop for price or were not given the tools for assessing a supplier's quality. Management had to develop long-term relationships with vendors, work with vendors to improve and maintain quality, train its own purchasing department in statistical quality control, require statistical evidence of quality from vendors, and insist that specifications be complete, including an understanding of how the material actually worked in manufacturing. Once management had changed purchasing systems and procedures, buyers could then not only be expected but also able to do their job in a new way. When top management had seriously committed to quality, lower-level personnel would be more likely to take action on problems that were within their control.

Accordingly, Deming delineated two means of process improvement: changing the "common causes" that were systemic (and were thus shared by numerous operators, machines, or products) and removing the "special causes" that produced nonrandom variation within systems (and were usually confined to individual employees or activities). Common causes included poor product design, incoming materials unsuited to their use, machines out of order, improper bills of materials, machinery that would not hold tolerances, poor physical conditions, and so on. Special causes included lack of knowledge or skill, worker inattention, or a poor lot of incoming materials. Management was responsible for common causes, and operators were responsible for special causes:

> The discovery of a special cause of variation and its removal are usually the responsibility of someone who is connected directly with some operation. . . . In contrast, there are common causes of defectives, of errors, of low rates of production, of low sales, of accidents. These are the responsibility of management. . . . The worker at a machine can

---

2. Jeremy Main, "The Curmudgeon Who Talks Tough on Quality," *Fortune*, June 25, 1984, p. 122.

3. Deming, *Quality*, p. ii.

4. Ibid., p. 1.

Figure A   A Typical Control Chart

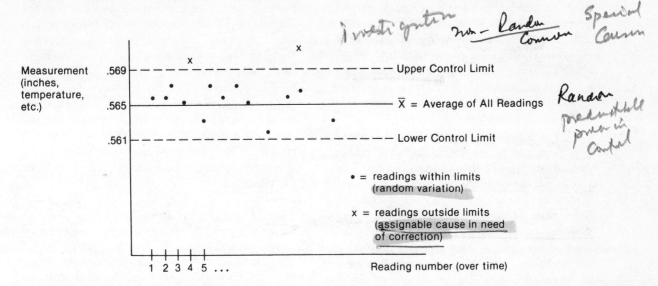

Handwritten margin notes: Investigation   non-Random Common   Special Causes

Measurement (inches, temperature, etc.)

.569 — — — — — — — — — — — — — — — Upper Control Limit

.565 ——————————————— $\overline{X}$ = Average of All Readings

Handwritten: Random predictable power in Control

.561 — — — — — — — — — — — — — — — Lower Control Limit

• = readings within limits (random variation)

x = readings outside limits (assignable cause in need of correction)

1 2 3 4 5 . . .   Reading number (over time)

do nothing about causes common to all machines. . . . He cannot do anything about the light; he does not purchase raw materials; the training, supervision, and the company's policies are not his.[5]

The key tool that Deming advocated to distinguish between systemic and special causes—and indeed, the key to quality management in general—was statistical process control (SPC). Developed by Walter Shewart while at Bell Labs in the 1930s and later refined by Deming in a well-known paper, "On the Statistical Theory of Errors," SPC was required because variation was an inevitable fact of industrial life. It was unlikely that two parts, even when produced by the same operator at the same machine, would ever be identical. The issue, therefore, was distinguishing acceptable variation from variation that could indicate problems. The rules of statistical probability provided a method for making this distinction.

Probability rules could determine whether variation was random or not, that is, whether it was due to chance. Random variation occurred within statistically determined limits. If variation remained within those limits, the process was a stable one and in control. As long as nothing changed the process, future variation

could be predicted easily, for it would remain indefinitely within the same statistical limits.

Data of this sort were normally collected and plotted on control charts kept by the operators themselves. Such charts graphically plotted actual performance readings (e.g., the outside diameters of pistons) on graphs that also depicted the upper and lower control limits for that characteristic, which were statistically determined (see *Figure A*).

As long as the readings, taken on a small sample of units at predetermined intervals (such as every half hour), fell between limits or did not show a trend or "run," the process was in control and no intervention was required, despite the obvious variation in readings. Readings that either fell outside the limits or produced a run indicated a problem to be investigated.

The practical value of distinguishing random from nonrandom variation was enormous. Operators now knew when to intervene in a process and when to leave it alone. Further, because readings were taken during the production process itself, unacceptable variation showed up early enough for corrective action, rather than after the fact.

Once a process was in control, readings that fell outside the limits indicated a special cause. When the cause of such nonrandom variation

5. Deming, *Quality*, p. 116.

was found and removed, the system returned to its stable state. For example, if a particular lot of goods showed yields that were below control limits, further analysis might determine that raw materials peculiar to that lot were the cause. The removal of such special causes, however, did not improve the system (i.e., raise yield levels), but simply brought it back under control at the preexisting yield.

To improve the system itself, common causes had to be removed. Simply because a system was in statistical control did not mean it was as good as it could be. Indeed, a process in control could produce a high proportion of defects. Control limits indicated what the process was, not what it should or could be. To move the average (yield, sales, defects, returns, etc.) up or down—and thus also move the control limits up or down—typically required the concerted efforts of engineering, research, sales, manufacturing, and other departments. To narrow the range of variation around the target point could consume even more effort. In both instances, control charts would readily document the improvements in the process.

Deming viewed training in the use of control charts as essential if workers were to know what constituted acceptable work. He was adamant that quotas, piecework, and numerical goals be eliminated. Instead, workers had to be shown good work and given the tools to do it. Such tools would also allow them to monitor their own work and correct it in real time, rather than find out about problems days or weeks later.

Control charts were but one part of the statistical approach to quality. Because 100% testing was inefficient, sampling techniques had been developed to provide a scientific basis on which to accept or reject production lots based on a limited number of units. Although sampling and control charts could indicate problems, they could not by themselves identify their causes. For that purpose, other statistical techniques were needed, such as Pareto analysis, Ishikawa or "fishbone" cause-and-effect diagrams, histograms, flow charts, and scatter diagrams.

By 1986 Deming's lectures concentrated more on management than SPC, but SPC re-

mained at the core of his approach. Many U.S. firms sought him out, and some, such as the Ford Motor Company, adopted his approach throughout the company with great success. Deming, who still worked out of the basement of his home with his secretary of 30 years, was hardly sanguine about the prospects for American business. He believed that it would take 30 years for Americans to match the progress of the Japanese and that the United States was still falling behind. With the specter of a lower U.S. standard of living, he concluded, "We should be pretty scared."[6]

Juran

Joseph M. Juran's impact on Japanese quality was usually considered second only to Deming's. At 82, he had enjoyed a varied and distinguished career that included stints as a business executive, government administrator, lecturer, writer, and consultant. After years of independent activity, he established the Juran Institute in 1979 to serve as a base for the seminars, consulting, conferences, and videotapes long associated with his name. His clientele included Texas Instruments, Du Pont, Monsanto, and Xerox.

Juran defined quality as "fitness for use," meaning that the users of a product or service should be able to count on it for what they needed or wanted to do with it. For example, a manufacturer should be able to process a purchased material or component to meet the demands of its customers while achieving high yields and minimal downtime in production; a wholesaler should receive a correctly labeled product, free from damage during shipment and easy to handle and display; and a consumer should receive a product that performed as claimed and did not break down—or, if it did, receive prompt and courteous adjustment of the claim.

Fitness for use had five major dimensions: quality of design, quality of conformance, availability, safety, and field use.[7] Quality of design was what distinguished a Rolls Royce from a Chevrolet and involved the design con-

---

6. Main, "The Curmudgeon Who Talks Tough on Quality," p. 122.

7. The key parameters of fitness for use, as well as their dimensions, vary somewhat in Juran's writings over a 35-year period. Their comprehensiveness and their spanning the entire product life cycle, however, are constants. The present discussion draws most heavily on Joseph M. Juran and Frank M. Gryna, Jr., *Quality Planning and Analysis* (New York: McGraw-Hill, 1980).

cept and its specification. Quality of conformance reflected the match between actual product and design intent and was affected by process choices, ability to hold tolerances, workforce training and supervision, and adherence to test programs. Availability referred to a product's freedom from disruptive problems and reflected both reliability (the frequency or probability of failure) and maintainability (the speed or ease of repair). Safety could be assessed by calculating the risk of injury due to product hazards. Field use referred to a product's conformance and condition after it reached customers' hands and was affected by packaging, transportation, storage, and field-service competence and promptness.

To achieve fitness for use, Juran developed a comprehensive approach to quality that spanned a product's entire life—from design through vendor relations, process development, manufacturing control, inspection and test, distribution, customer relations, and field service. Each area was carefully dissected, and approaches were proposed to specify and quantify its impact on the various elements of fitness for use. A broad range of statistical techniques was included to assist in the analysis.

Juran's approach to reliability provides a representative example. His reliability program began by establishing reliability goals. It then apportioned these among product components; identified critical components; identified possible modes, effects, and causes of failures; and developed solutions for those most critical to successful product operation and safety. Juran also discussed the setting of realistic tolerances, design reviews, vendor selection, and testing of designs. Statistical methods for improving reliability included analysis of various types of failure rates, analysis of relationships between component and system reliability, and setting of tolerance limits for interacting dimensions. The aims of these activities were quantified reliability goals, a systematic guide for achieving them, and a measurement and monitoring system for knowing when they had been achieved.

Although Juran's analytical methods could identify areas needing improvement and could help make and track changes, they were in the language of the shop floor: defect rates, failure modes, not within specification, and the like.

Juran recognized that such measures were not likely to attract top management attention; for this reason, he advocated a cost-of-quality (COQ) accounting system. Such a system spoke top management's language—money. Quality costs were costs "associated solely with defective product—the costs of making, finding, repairing, or avoiding defects."[8] They were of four types: internal failure costs (from defects discovered before shipment); external failure costs (from defects discovered after shipment); appraisal costs (for assessing the condition of materials and product); and prevention costs (for keeping defects from occurring in the first place). (See *Exhibit 2*.) In most companies, external and internal failure costs together accounted for 50% to 80% of COQ. When these were converted to dollars or presented as a percentage of sales or profits, top management usually took notice.

COQ not only provided management with a dollar cost for defective products, it also established the goal of quality programs: to keep improving quality until there was no longer a positive economic return. This occurred when the total costs of quality were minimized (see *Exhibit 3*). Two assumptions were built into this analysis: that failure costs approached zero as defects became fewer and fewer, and that prevention and appraisal costs together approached infinity as defects were reduced to lower and lower levels. COQ minimization therefore occurred at the point where additional spending on prevention and appraisal was no longer justified because it produced smaller savings in failure costs.

This approach had important practical implications. It implied that zero defects was not a practical goal, for to reach that level prevention and appraisal costs would have to rise so substantially that total costs of quality would not be minimized. As long as prevention and appraisal costs were cheaper (on a per-unit basis) than failure costs, Juran argued, resources should continue to go to prevention and testing. When prevention activities started to pull COQ unit costs up rather than down, however, it was time to maintain quality rather than attempt to reduce it further.

To reach and maintain this minimum cost of quality, Juran proposed a three-pronged ap-

---

8.   Juran and Gryna, *Quality Planning*, p. 13.

Figure B   Juran's Sporadic and Chronic Problems

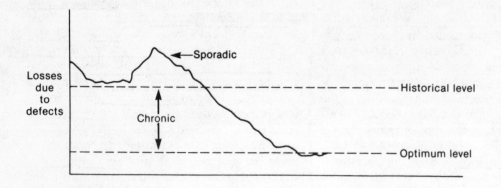

Note: This figure is adapted from Juran and Gryna, *Quality Planning,* p. 100.

proach: breakthrough projects, the control sequence, and annual quality programs. In the early stages, when a firm's failure costs greatly exceeded its prevention and appraisal costs, there were significant opportunities for breakthrough projects, aimed at chronic problems. Problems, such as the need to revise tolerances, were ignored because they were neither dramatic nor thought to be solvable. The "breakthrough sequence" involved identifying the "vital few" projects, selling them to management, organizing to analyze the issues and to involve the key people who were needed for implementation, and overcoming resistance to change (see *Exhibit 4*). Juran claimed that most breakthrough analyses found that over 80% of the problems (e.g., defect rates, scrap rates) were under management control and fewer than 20% were caused by operators.

After successive breakthrough projects, a firm reached the point of optimal quality—in Juran's formulation, the bottom of the COQ curve. The organization then needed to employ the control sequence to preserve its gains. This sequence was actually a large feedback loop. The first step was to choose an objective to control, then to define a unit of measure, set a numerical standard or goal, create a means of measuring performance, and mobilize the organization to report the measurements. After these preparatory steps, an action cycle was repeated over and over: actual performance was compared with standard, and action was taken (if needed) to close the gap.[9]

The control sequence was also used to attack sporadic problems—sudden, usually dramatic changes in the status quo, such as a worn cutting tool. Eliminating sporadic problems only returned processes to their historical levels; to improve them to optimum levels, breakthrough teams were needed because chronic problems were involved. Juran's contrast between these two types of problems is illustrated in *Figure B*.

Both the control and breakthrough processes demanded sophisticated analysis and statistics. The comprehensiveness of Juran's program (it ran from vendor relations through customer service and covered all the functions in between) required high-level planning and coordination as well. For this reason, Juran argued that a new group of professionals—quality control engineers—was needed. This department would be involved in high-level quality planning, coordinating the activities of other departments, setting quality standards, and providing quality measurements. Juran also believed that top management had to give overall leadership and support to quality improvement for it to succeed.

Juran's major vehicle for top management involvement was the annual quality program. Akin to long-range financial planning and the annual budget process, this program gave top management quality objectives and was especially important for internalizing the habit of

---

9.   This description of the control sequence is based on Joseph M. Juran, *Managerial Breakthrough* (New York: McGraw-Hill, 1964), pp. 183–187.

quality improvement to ensure that complacency did not set in.

Crosby

[handwritten: Quality = Conformance to requirement]

Philip B. Crosby started in industry as an inspector. He eventually rose through the ranks at several companies to become vice president of quality at ITT. In 1979 he left ITT to found Philip Crosby Associates, Inc., along with the Crosby Quality College, which by 1986 approximately 35,000 executives and managers had attended. General Motors owned over 10% of Crosby stock and had set up its own Crosby school, as had IBM, Johnson & Johnson, and Chrysler.

Crosby directed his message to top managers. He sought to change their perceptions and attitudes about quality. Typically, top managers viewed quality as intangible or else to be found only in high-end products. Crosby, however, spoke of quality as "conformance to requirements" and believed that any product that consistently reproduced its design specifications was of high quality. In this sense, a Pinto that met Pinto requirements was as much a quality product as a Cadillac that conformed to Cadillac requirements.

American managers must pursue quality to help them compete, Crosby argued. In fact, he believed that if quality were improved, total costs would inevitably fall, allowing companies to increase profitability. This reasoning led to Crosby's most famous claim—that quality was "free."[10]

[handwritten margin note: Zero def. 1960 at Martin]

Ultimately, the goal of quality improvement was zero defects, to be achieved through prevention rather than after-the-fact inspection. Crosby had popularized the zero defects movement, but it had actually originated in the United States at the Martin Company in the 1960s, where Crosby was employed. The company had promised and delivered a perfect missile, with limited inspection and rework, and its managers had concluded that perfection was possible if, in fact, it was expected. The company then developed a philosophy and program to support that goal.

Crosby elaborated on this approach. He believed that the key to quality improvement was changing top management's thinking. If management expected imperfection and defects, it would get them, for workers would bring similar expectations to their jobs. But if management established a higher standard of performance and communicated it thoroughly to all levels of the company, zero defects was possible. Thus, according to Crosby, zero defects was a management standard and not simply a motivational program for employees.

To help managers understand the seriousness of their quality problems, Crosby provided two primary tools: cost of quality measures and the management maturity grid (see *Exhibit 5*). Costs of quality, which Crosby estimated to be between 15% and 20% of sales at most companies, were useful for showing top management the size of its quality problem and the opportunities for profitable improvement. The management maturity grid was used for self-assessment. It identified five states of quality awareness: uncertainty (the company failed to recognize quality as a management tool); awakening (quality was recognized as important, but management put off taking action); enlightenment (management openly faced and addressed quality problems by establishing a formal quality program); wisdom (prevention was working well, problems were identified early, and corrective action was routinely pursued); and certainty (quality management was an essential part of the company, and problems occurred only infrequently). For each of these five stages, Crosby also examined the status of the quality organization, problem-handling procedures, reported and actual costs of quality as percentages of sales, and quality improvement actions.

[handwritten margin note: 5 State • Uncertainty • awakening • enlightenment • Wisdom • Certainty]

Once companies had positioned themselves on the management maturity grid, Crosby offered a 14-point program for quality improvement (see *Exhibit 6*). It emphasized prevention over detection, and focused on changing corporate culture rather than on analytical or statistical tools. The program was designed as a guide for securing management commitment and gaining employees' involvement through actions such as Zero Defects Day. Crosby believed every company should tailor its own defect-prevention program; nevertheless, the goal should always be zero defects. In this process top management played a leadership role; quality professionals played a modest but important role as facilitators, coordinators, trainers, and technical assistants, and hourly workers were secondary.

---

10. Philip B. Crosby, *Quality Is Free* (New York: McGraw-Hill, 1979).

Exhibit 1   Deming's 14 Points

---

1. **Create constancy of purpose for improvement of product and service.**[a] Management must change from a preoccupation with the short run to building for the long run. This requires dedication to innovation in all areas to best meet the needs of customers.

2. **Adopt the new philosophy.** Shoddy materials, poor workmanship, defective products, and lax service must become unacceptable.

3. **Cease dependence on mass inspection.** Inspection is equivalent to planning for defects; it comes too late and is ineffective and costly. Instead, processes must be improved.

4. **End the practice of awarding business on price tag alone.** Price has no meaning without a measure of the quality being purchased. Therefore, the job of purchasing will change only after management establishes new guidelines. Companies must develop long-term relationships and work with fewer suppliers. Purchasing must be given statistical tools to judge the quality of vendors and purchased parts. Both purchasing and vendors must understand specifications, but they must also know how the material is to be used in production and by the final customer.

5. **Constantly and forever improve the system of production and service.** Waste must be reduced and quality improved in every activity: procurement, transportation, engineering, methods, maintenance, sales, distribution, accounting, payroll, customer service, and manufacturing. Improvement, however, does not come from studying the defects produced by a process that is in control but from studying the process itself. Most of the responsibility for process improvement rests with management.

6. **Institute modern methods of training on the job.** Training must be restructured and centered on clearly defined concepts of acceptable work. Statistical methods must be used for deciding when training has been completed successfully.

7. **Institute modern methods of supervising.** Supervisors must be empowered to inform upper management about conditions that need correction; once informed, management must take action. Barriers that prevent hourly workers from doing their jobs with pride must be removed.

8. **Drive out fear.** Because of the tremendous economic losses caused by fear on the job, people must not be afraid to ask questions, to report problems, or to express ideas.

9. **Break down barriers between departments.** Members of the research, design, procurement, sales, and receiving departments must learn about problems with raw materials and specifications in production and assembly. Each discipline must stop optimizing its own work and instead work together as a team for the company as a whole. Multidisciplinary quality-control circles can help improve design, service, quality, and costs.

10. **Eliminate numerical goals for the work force.** Targets, slogans, pictures, and posters urging people to increase productivity must be eliminated. Most of the necessary changes are out of workers' control, so such exhortations merely cause resentment. Although workers should not be given numerical goals, the company itself must have a goal: never-ending improvement.

11. **Eliminate work standards and numerical quotas.** Quotas focus on quantity, not quality. Therefore, work standards practically guarantee poor quality and high costs. Work standards that state percentage-defective or scrap goals normally reach those targets but never exceed them. Piecework is even worse, for it pays people for building defective units. But if someone's pay is docked for defective units, that is unfair, for the worker did not create the defects.

12. **Remove barriers that hinder the hourly workers.** Any barrier that hinders pride in work must be removed, including not knowing what good work is, supervisors motivated by quotas, off-gauge parts and material, and no response to reports of out-of-order machines.

13. **Institute a vigorous program of education and training.** Because quality and productivity improvements change the number of people needed in some areas and the jobs required, people must be continually trained and retrained. All training must include basic statistical techniques.

14. **Create a structure in top management that will push every day on the above 13 points.**

---

a. Deming's words are in bold headings . The remainder of each paragraph paraphrases his discussions.

Exhibit 2   Juran's Categories of Quality Costs

Internal failure costs = costs from product defects before shipment to the customer. They include the following:

- *Scrap* – net losses in labor and material resulting from defective goods that cannot economically be repaired or used.
- *Rework* – costs of correcting defective products to make them usable.
- *Retest* – costs of reinspection and retesting of products that have been reworked.
- *Downtime* – costs of idle facilities, equipment, and labor due to defective products.
- *Yield losses* – costs of process yields lower than could be attained through improved process control.
- *Disposition* – the time of those involved in determining whether nonconforming products are usable and what should be done with them.

External failure costs = costs associated with defects found after shipment to customer. They include the following:

- *Complaint adjustment* – costs of investigating and responding to complaints due to defective products, faulty installation, or improper instructions to users.
- *Returned material* – costs associated with receiving and replacing defective products returned from the field.
- *Warranty charges* – costs of services and repairs performed under warranty contracts.
- *Allowances* – income losses due to downgrading products for sale as seconds and to concessions made to customers who accept substandard products as is.

Appraisal costs = costs associated with discovering the condition of products and raw materials. They include the following:

- *Incoming materials inspection* – costs associated with determining the quality of vendors' products.
- *Inspection and test* – costs of checking product conformance throughout design and manufacture, including tests done on customers' premises.
- *Maintaining accuracy of test equipment* – costs of operating and maintaining measuring instruments.
- *Materials and services consumed* – costs of products consumed in destructive tests; also materials and services (e.g., electric power) consumed in testing.
- *Evaluation of stocks* – costs of testing products in storage to assess their condition.

Prevention costs = costs associated with preventing defects and limiting failure and appraisal costs. They include the following:

- *Quality planning* – costs of creating and communicating plans and data systems for quality, inspection, reliability, and related activities—includes the costs of preparing all necessary manuals and procedures.
- *New products review* – costs of preparing bid proposals, evaluating new designs, preparing test and experimentation programs, and related quality activities associated with launching new products.
- *Training* – costs of developing and conducting training programs aimed at improving quality performance.
- *Process control* – costs of process control aimed at achieving fitness for use, as distinguished from productivity (a difficult distinction to make in practice).
- *Quality data acquisition and analysis* – costs of operating the quality data system to get continuing data on quality performance.
- *Quality reporting* – costs of bringing together and presenting quality data to upper management.
- *Improvement projects* – costs of building and implementing breakthrough projects.

Note: This is a summary and rewording of Juran and Gryna, *Quality Planning*, pp. 14–16.

Exhibit 3 Minimizing the Costs of Quality

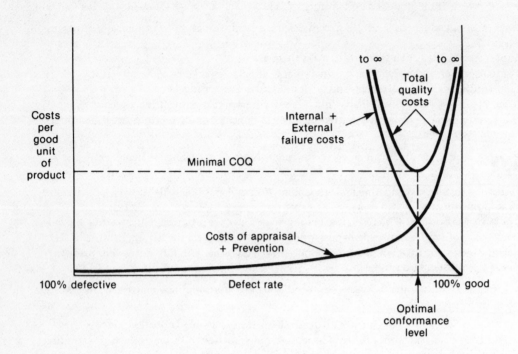

Note: This figure is adapted from Juran and Gryna, *Quality Planning*, p. 27.

Exhibit 4   Juran's Breakthrough Sequence

1. **Breakthrough in attitudes.** Managers must first prove that a breakthrough is needed and then create a climate conducive to change. To demonstrate need, data must be collected to show the extent of the problem; the data most convincing to top management are usually cost-of-quality figures. To get the resources required for improvement, expected benefits can be monetized and presented in terms of return on investment.

2. **Identify the vital few projects.** Pareto analysis is used to distinguish the vital few projects from the trivial many and to set priorities based on problem frequency.

3. **Organize for breakthrough in knowledge.** Two organizational entities should be established—a steering group and a diagnostic group. The steering group, composed of people from several departments, defines the program, suggests possible problem causes, gives the authority to experiment, helps overcome resistance to change, and implements the solution. The diagnostic group, composed of quality professionals and sometimes line managers, is responsible for analyzing the problem.

4. **Conduct the analysis.** The diagnostic group studies symptoms, develops hypotheses, and experiments to find the problem's true causes. It also tries to determine whether defects are primarily operator controllable or management controllable. (A defect is operator controllable only if it meets three criteria: operators know what they are supposed to do, have the data to understand what they are actually doing, and are able to regulate their own performance.) Theories can be tested by using past data and current production data and by conducting experiments. With this information, the diagnostic group then proposes solutions to the problem.

5. **Determine how to overcome resistance to change.** The need for change must be established in terms that are important to the key people involved. Logical arguments alone are insufficient. Participation is therefore required in both the technical and social aspects of change.

6. **Institute the change.** Departments that must take corrective action must be convinced to cooperate. Presentations to these departments should include the size of the problem, alternative solutions, the cost of recommended changes, expected benefits, and efforts taken to anticipate the change's impact on employees. Time for reflection may be needed, and adequate training is essential.

7. **Institute controls.** Controls must be set up to monitor the solution and see that it works and to keep abreast of unforeseen developments. Formal follow-up is provided by the control sequence used to monitor and correct sporadic problems.

Note: This summary is adapted from Juran and Gryna, *Quality Planning*, pp. 100–129, and Juran, *Managerial Breakthrough*, pp. 15–17.

Exhibit 5   Crosby's Quality Management Maturity Grid

Measurement Categories	Stage I: Uncertainty	Stage II: Awakening	Stage III: Enlightenment	Stage IV: Wisdom	Stage V: Certainty
Management under-standing and attitude	Fails to see quality as a management tool.	Supports quality management in theory but is unwilling to provide the necessary money or time.	Learns about quality management and becomes supportive.	Participates personally in quality activities.	Regards quality management as essential to the company's success.
Quality organization status	Quality activities are limited to the manufacturing or engineering department and are largely appraisal and sorting.	A strong quality leader has been appointed, but quality activities remain focused on appraisal and sorting and are still limited to manufacturing and engineering.	Quality department reports to top management, and its leader is active in company management.	Quality manager is an officer of the company. Prevention activities have become important.	Quality manager is on the board of directors. Prevention is the main quality activity.
Problem handling	Problems are fought as they occur and are seldom fully resolved; "fire-fighting" dominates.	Teams are established to attack major problems, but the approach remains short term.	Problems are resolved in an orderly fashion, and corrective action is a regular event.	Problems are identified early in their development.	Except in the most unusual cases, problems are prevented.
Cost of quality as percentage of sales	Reported: unknown Actual: 20%	Reported: 5% Actual: 18%	Reported: 8% Actual: 12%	Reported: 6.5% Actual: 8%	Reported: 2.5% Actual: 2.5%
Quality improvement actions	No organized activities.	Activities are motivational and short term.	Implements the 14-step program with full understanding.	Continues the 14-step program and starts Make Certain.	Quality improvement is a regular and continuing activity.
Summation of company quality posture	"We don't know why we have quality problems."	"Must we always have quality problems?"	"Because of management commitment and quality improvement programs, we are identifying and resolving our quality problems."	"We routinely prevent defects from occurring."	"We know why we don't have quality problems."

Note: This chart is adapted from Crosby, *Quality Is Free*, pp. 32–33.

Exhibit 6 Crosby's 14-Point Program

1. **Management commitment.** Top management must become convinced of the need for quality improvement and must make its commitment clear to the entire company. This should be accompanied by a written quality policy, stating that each person is expected to "perform exactly like the requirement, or cause the requirement to be officially changed to what we and the customers really need."

2. **Quality improvement team.** Management must form a team of department heads (or those who can speak for their departments) to oversee quality improvement. The team's role is to see that needed actions take place in its departments and in the company as a whole.

3. **Quality measurement.** Quality measures that are appropriate to every activity must be established to identify areas needing improvement. In accounting, for example, one measure might be the percentage of late reports; in engineering, the accuracy of drawings; in purchasing, rejections due to incomplete descriptions; and in plant engineering, time lost because of equipment failures.

4. **Cost of quality evaluation.** The controller's office should make an estimate of the costs of quality to identify areas where quality improvements would be profitable.

5. **Quality awareness.** Quality awareness must be raised among employees. They must understand the importance of product conformance and the costs of nonconformance. These messages should be carried by supervisors (after they have been trained) and through such media as films, booklets, and posters.

6. **Corrective action.** Opportunities for correction are generated by Steps 3 and 4, as well as by discussions among employees. These ideas should be brought to the supervisory level and resolved there, if possible. They should be pushed up further if that is necessary to get action.

7. **Zero defects planning.** An ad hoc zero defects committee should be formed from members of the quality improvement team. This committee should start planning a zero defects program appropriate to the company and its culture.

8. **Supervisor training.** Early in the process, all levels of management must be trained to implement their part of the quality improvement program.

9. **Zero Defects Day.** A Zero Defects Day should be scheduled to signal to employees that the company has a new performance standard.

10. **Goal setting.** To turn commitments into action, individuals must establish improvement goals for themselves and their groups. Supervisors should meet with their people and ask them to set goals that are specific and measurable. Goal lines should be posted in each area and meetings held to discuss progress.

11. **Error cause removal.** Employees should be encouraged to inform management of any problems that prevent them from performing error-free work. Employees need not do anything about these problems themselves; they should simply report them. Reported problems must then be acknowledged by management within 24 hours.

12. **Recognition.** Public, nonfinancial appreciation must be given to those who meet their quality goals or perform outstandingly.

13. **Quality councils.** Quality professionals and team chairpersons should meet regularly to share experiences, problems, and ideas.

14. **Do it all over again.** To emphasize the never-ending process of quality improvement, the program (Steps 1–13) must be repeated. This renews the commitment of old employees and brings new ones into the process.

Note: This summary is adapted from Crosby, *Quality Is Free*, pp. 132–139, 175–259.

in·teg·ri·ty \ in-'teg-rət-ē \ n (15c) 1: an unimpaired condition: SOUNDNESS 2: firm adherence to a code of esp. moral or artistic values: INCORRUPTIBILITY 3: the quality or state of being complete or undivided: COMPLETENESS

The Power of Product Integrity

Kim B. Clark and Takahiro Fujimoto

Some companies consistently develop products that succeed with customers. Other companies often fall short. What differentiates them is integrity. Every product reflects the organization and the development process that created it. Companies that consistently develop successful products–products with integrity–are themselves coherent and integrated. Moreover, this coherence is distinguishable not just at the level of structure and strategy but also, and more important, at the level of day-to-day work and individual understanding. Companies with organizational integrity possess a source of competitive advantage that rivals cannot easily match.

The primacy of integrity, in products and organizations alike, begins with the role new products play in industrial competition and with the difficulty of competing on performance or price alone. New products have always fascinated and excited customers, of course. Henry Ford's Model A made front-page news after near-riots erupted outside dealers' showrooms. But today, in industries ranging from cars and computers to jet engines and industrial con-

Kim B. Clark is the Harry E. Figgie, Jr. Professor of Business Administration at the Harvard Business School. His most recent article in HBR was "What Strategy Can Do for Technology" (November-December 1989). Takahiro Fujimoto is assistant professor of business administration at Tokyo University. Their new book, Product Development Performance, *will be published in early 1991 by the Harvard Business School Press.*

Product Integrity ?

trols, new products are the focal point of competition. Developing high-quality products faster, more efficiently, and more effectively tops the competitive agenda for senior managers around the world.

Three familiar forces explain why product development has become so important. In the last two decades, intense international competition, rapid technological advances, and sophisticated, demanding customers have made "good enough" unsatisfactory in more and more consumer and industrial markets. Yet the very same forces are also making product integrity harder and harder to achieve.

Consider what happened when Mazda and Honda each introduced four-wheel steering to the Japanese auto market in 1987. Although the two steering systems used different technologies – Mazda's was based

When Mazda put racy, four-wheel steering in a conservative family car, potential customers felt the mismatch.

on electronic control, while Honda's was mechanical – they were equally sophisticated, economical, and reliable. Ten years earlier, both versions probably would have met with success. No longer. A majority of Honda's customers chose to install four-wheel steering in their new cars; Mazda's system sold poorly and was widely regarded as a failure.

Why did consumers respond so differently? Product integrity. Honda put its four-wheel steering system into the Prelude, a two-door coupe with a sporty, progressive image that matched consumers' ideas about the technology. The product's concept and the new component fit together seamlessly; the car sent a coherent message to its potential purchasers. In contrast, Mazda introduced its four-wheel steering system in the 626, a five-door hatchback that consumers associated with safety and dependability. The result was a mismatch between the car's conservative, family image and its racy steering system. Too sophisticated to be swayed by technology alone (as might have been the case a decade before), Mazda's potential customers saw no reason to buy a car that did not satisfy their expectations in every respect, including "feel." (Mazda's new advertising slogan, "It just feels right," suggests the company's managers took this lesson to heart.)

Product integrity is much broader than basic functionality or technical performance. Customers who have accumulated experience with a product expect new models to balance basic functions and economy with more subtle characteristics. Consumers expect new products to harmonize with their values and lifestyles. Industrial customers expect them to mesh with existing components in a work system or a production process. The extent to which a new product achieves this balance is a measure of its integrity. (One of integrity's primary metrics is market share, which reflects how well a product attracts and satisfies customers over time.)

Product integrity has both an internal and an external dimension. Internal integrity refers to the consistency between a product's function and its structure: the parts fit smoothly, the components match and work well together, the layout maximizes the available space. Organizationally, internal integrity is achieved mainly through cross-functional coordination within the company and with suppliers. Efforts to enhance internal integrity through this kind of coordination have become standard practice among product developers in recent years.

External integrity refers to the consistency between a product's performance and customers' expectations. In turbulent markets like those in which Honda and Mazda were competing, external integrity is critical to a new product's competitiveness. Yet for the most part, external integrity is an underexploited opportunity. Companies assign responsibility for anticipating what customers will want to one or more functional groups (the product planners in marketing, for example, or the testers in product engineering). But they give little or no attention to integrating a clear sense of customer expectations into the work of the product development organization as a whole.

Of course, there are exceptions. In a six-year study of new product development (see the insert "Focus on Development"), we found a handful of companies that consistently created products with integrity. What set these companies apart was their seamless pattern of organization and management. The way people did their jobs, the way decisions were made, the way suppliers were integrated into the company's own efforts – everything cohered and supported company strategy. If keeping the product line fresh and varied was a goal, speed and flexibility were apparent at every step in the development process, as were the habits and assumptions that accustom people and organizations to being flexible and to solving problems quickly. For example, product plans relied on large numbers of parts from suppliers who focused on meeting tight schedules and high quality standards even when designs changed late in the day. Product and process engineers jointly developed body panels and the dies to make them through informal, intense interactions that cut out unnecessary mistakes and solved problems on the spot. Production people built high-quality prototypes that tested the design against the realities of commercial production early in the game and so eliminated expensive delays and rework later on.

The examples we draw on in this article all come from the auto industry. We chose to look at a single industry worldwide so that we could identify the factors that separate outstanding performers from competitors making similar products for similar markets around the globe. But our basic findings apply to businesses as diverse as semiconductors, soup, and commercial construction. Wherever managers face a turbulent, intensely competitive market, product integrity – and the capacity to create it – can provide a sustainable competitive advantage.

The Power of a Product Concept

Products are tangible objects – things you can see, touch, and use. Yet the process of developing new products depends as much on the flow of information as it does on the flow of materials. Consider how a new product starts and ends.

Before a customer unpacks a new laptop computer or sets up a high-speed packaging machine, and long before a new car rolls off the showroom floor, the product (or some early version of it) begins as an idea. Next, that idea is embodied in progressively more detailed and concrete forms: ideas turn into designs, designs into drawings, drawings into blueprints, blueprints into prototypes, and so on until a finished product emerges from the factory. When it is finally in customers' hands, the product is converted into information once again.

If this last statement sounds odd, think about what actually happens when a potential buyer test-drives a new car. Seated behind the wheel, the customer receives a barrage of messages about the vehicle's performance. Some of these messages are delivered directly by the car: the feel of the acceleration, the responsiveness of the steering system, the noise of the engine, the heft of a door. Others come indirectly: the look on people's faces as the car goes by, comments from passengers, the driver's recollection of the car's advertising campaign. All these messages influence the customer's evaluation, which will largely depend on how he or she interprets them. In essence, the customer is consuming the product *experience*, not the physical product itself.

Developing this experience – and the car that will embody it – begins with the creation of a product concept. A powerful product concept specifies how the new car's basic functions, structures, and messages will attract and satisfy its target customers. In sum, it defines the character of the product from a customer's perspective.

The phrase "pocket rocket," for example, captures the basic concept for a sporty version of a subcompact car. Small, light, and fast, a pocket rocket should also have quick, responsive handling and an aggressive design. While the car should sell at a premium compared with the base model, it should still be affordable. And the driving experience should be fun: quick at the getaway, nimble in the turns, and very fast on the straightaways. Many other design and engineering details would need definition, of course, for the car to achieve its objectives. But the basic concept of an affordable and fun-to-drive pocket rocket would be critical in guiding and focusing creative ideas and decisions.

By definition, product concepts are elusive and equivocal. So it is not surprising that when key project participants are asked to relate the concept for a new vehicle, four divergent notions of value emerge. Those for whom the product concept means *what the product does* will couch their description in

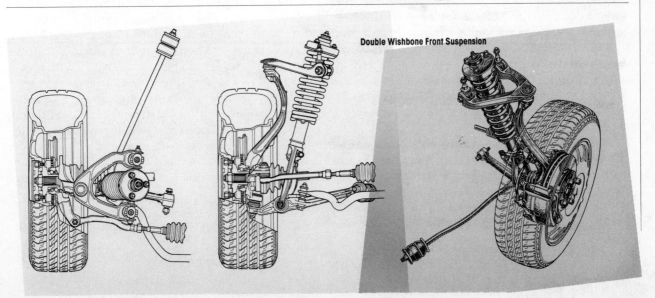

Double Wishbone Front Suspension

Focus on Development

What are the sources of superior performance in product development? What accounts for the wide differences in performance among companies in the same industry? To answer those questions, we studied 29 major development projects in 20 automobile companies around the world. (Three companies are headquartered in the United States, eight in Japan, and nine in Europe.) The projects ranged from micro-mini cars and small vans to large luxury sedans, with suggested retail prices from $4,300 to more than $40,000. Our research methods included structured and unstructured interviews, questionnaires, and statistical analysis. Throughout the study, we strove to develop a consistent set of data (including both measures of performance and patterns of organization and management) so that we could identify the constants among projects that differed greatly in scope and complexity.

We chose to concentrate on the automobile industry because it is a microcosm of the new industrial competition. In 1970, a handful of auto companies competed on a global scale with products for every market segment; today more than 20 do. Customers have grown more discerning, sophisticated, and demanding. The number of models has multiplied, even as growth has slowed, and technology is ever-more complex and diverse. In 1970, for example, the traditional V-8 engine with 3-speed automatic transmission and rear-wheel drive was the technology of choice for 80% of the cars produced in the United States. By the early 1980s, consumers could choose among 34 alternative configurations. In this environment, fast, efficient, effective product development has become the focal point of competition and managerial action.

terms of performance and technical functions. Others, for whom the concept means *what the product is*, will describe the car's packaging, configuration, and main component technologies. Others, for whom product concept is synonymous with *whom the product serves*, will describe target customers. Still others, reflecting their interpretation of the concept as *what the product means to customers*, will respond thematically, describing the car's character, personality, image, and feel.

The most powerful product concepts include all these dimensions. They are often presented as images or metaphors (like pocket rocket) that can evoke many different aspects of the new product's message without compromising its essential meaning. Honda Motor is one of the few auto companies that make the generation of a strong product concept the first step in their development process.

When Honda's engineers began to design the third-generation (or 1986) Accord in the early 1980s, they did not start with a sketch of a car. The engineers started with a concept—"man maximum, machine minimum"—that captured in a short, evocative phrase the way they wanted customers to feel about the car. The concept and the car have been remarkably successful: since 1982, the Accord has been one of the best-selling cars in the United States; in 1989, it was the top-selling car. Yet when it was time to design the 1990 Accord, Honda listened to the market, not to its own success. Market trends were indicating a shift away from sporty sedans toward family models. To satisfy future customers' expectations —and to reposition the Accord, moving it up-market just a bit—the 1990 model would have to send a new set of product messages.

As the first step in developing an integrated product concept, the Accord's project manager (the term Honda uses is "large product leader") led a series of small group discussions involving close to 100 people in all. These early brainstorming sessions involved people from many parts of the organization, including body engineering, chassis engineering, interior design, and exterior design. In line with Honda tradition, the groups developed two competing concepts in parallel. The subject of the discussions was abstract: what would be expected of a family sedan in the 1990s. Participants talked frequently about "adult taste" and "fashionability" and eventually came to a consensus on the message the new model would deliver to customers—"an adult sense of reliability." The ideal family car would allow the driver to transport family and friends with confidence, whatever the weather or road conditions; passengers would always feel safe and secure.

This message was still too abstract to guide the product and process engineers who would later be making concrete choices about the new Accord's specifications, parts, and manufacturing processes. So the next step was finding an image that would personify the car's message to consumers. The image the product leader and his team emerged with was "a rugby player in a business suit." It evoked rugged, physical contact, sportsmanship, and gentlemanly behavior—disparate qualities the new car would have to convey. The image was also concrete enough to translate clearly into design details. The decision to replace the old Accord's retractable headlamps with headlights made with a pioneering technology developed by Honda's supplier, Stanley, is a good example. To the designers and engineers, the new lights' totally transparent cover glass symbolized the will of a rugby player looking into the future calmly, with clear eyes.

The next and last step in creating the Accord's product concept was to break down the rugby player image into specific attributes the new car would have to possess. Five sets of key words captured what the product leader envisioned: "open-minded," "friendly communication," "tough spirit," "stress-free," and "love forever." Individually and as a whole, these key words reinforced the car's message to consumers. "Tough spirit" in a car, for example, meant maneuverability, power, and sure handling in extreme driving conditions, while "love forever" translated into long-term reliability and customer satisfaction. Throughout the course of the project, these phrases provided a kind of shorthand to help people make coherent design and hardware choices in the face of competing demands. Moreover, they were a powerful spur to innovation.

Consider this small slice of the process. To approximate the rugby player's reliability and composure ("stress-free"), the engineers had to eliminate all unnecessary stress from the car. In technical terms, this meant improving the car's NVH, or noise, vibration, and harshness characteristics. That, in turn, depended on reducing the "three gangs of noise," engine noise, wind noise, and road noise.

To reduce engine noise, the product engineers chose a newly developed balance shaft that rotated twice as fast as the engine and offset its vibration. The shaft made the Accord's compact 4-cylinder engine as quiet as a V-6 and conserved space in the process. But since the shaft was effective only when the engine was turning over reasonably quickly, the product engineers also had to design a new electrically controlled engine mount to minimize vibration when the engine was idling.

Moreover, once the engine was quieter, other sources of noise became apparent. The engineers learned that the floor was amplifying noise from the engine, as was the roof, which resonated with the engine's vibration and created unpleasant, low-frequency booming sounds. To solve these problems, the engineers inserted paper honeycomb structures 12

> **When it was time to redesign the Accord, Honda listened to the market, not to its own success.**

to 13 millimeters thick in the roof lining—a solution that also improved the roof's structural rigidity and contributed to the car's tough spirit. They also re-

designed the body floor, creating a new sandwich structure of asphalt and sheet steel, which similarly strengthened the body shell.

Multiply this example hundreds of times over and it is clear why a strong product concept is so important. At its core, the development process is a complex system for solving problems and making decisions. Product concepts like those developed at Honda give people a clear framework for finding solutions and making decisions that complement one another and ultimately contribute to product integrity.

Organizing for Integrity

When cars were designed and developed by a handful of engineers working under the direction of a Henry Ford, a Gottlieb Daimler, or a Kiichiro Toyoda, organization was not an issue. What mattered were the engineers' skills, the group's chemistry, and the master's guidance. These are still vital to product integrity; but the organizational challenge has become immeasurably more complex. Developing a new car involves hundreds (if not thousands) of people working on specialized pieces of the project in many different locations for months or even years at a time. Whether their efforts have integrity—whether the car performs superbly and delights customers—will depend on how the company organizes development and the nature of the leadership it creates.

Efforts to organize development effectively are rooted in the search for solutions to two basic problems. One is designing, building, and testing the product's parts and subsystems so that every element achieves a high level of performance. In a car, this means that the brakes hold on wet or icy roads, the suspension gives a smooth ride on rough roads, the car corners well on sharp turns, and so on. Because performance at this level is driven by expertise and deep understanding, some specialization, both for individuals and for the organization, is essential. Yet specialization is a double-edged sword. By complicating communication and coordination across the organization, it complicates the second problem that development organizations face: achieving product integrity.

When markets were relatively stable, product life cycles long, and customers concerned most with technical performance, companies could achieve product integrity through strong functional organizations. Managers could commit whatever resources and time it took to make products that worked well, and external integrity (matching the product to customer expectations) was simply a by-product of those efforts. But as competition intensified and customers' needs and wants grew harder to predict, integration became an explicit goal for most product developers. By the late 1980s, even the most resolutely functional development organizations had established formal mechanisms such as coordination committees, engineering liaisons, project managers, matrix structures, and cross-functional teams to improve product development.

Structural mechanisms like these are only a small part of achieving product integrity, however. At best—when they are reinforced and supported by the behaviors, attitudes, and skills of people in every part of the development organization—they speed problem solving and improve the quality of the solutions. But by design, they are focused inward; they do not address integrity's external dimension. So unless the company makes a deliberate effort to integrate customers into the development process, it is likely to create products that are fresh, technologically advanced, and provide good value but that often fall short with sophisticated consumers.

For this reason, external integration is the single most important task for new product development. It represents a conscious organizational effort to enhance the external integrity of the development process by matching the philosophy and details of product design to the expectations of target customers. Generating a distinctive product concept that anticipates future customers' needs and wants is the first step in external integration. Infusing this concept into drawings, plans, detailed designs, and, ultimately, the product itself is the substance of its ongoing work.

To get some sense of how thorough (and hard) this infusion process actually is, consider a few of the conflicts Honda faced during the planning stage for the third-generation Accord.

The vehicle's product concept (man maximum, machine minimum) included maximum space and visibility for the occupants, minimum space for the car's mechanisms, a wide, low body for aesthetics, superb handling and stability, and superior economy in operation. To convey a feeling of spaciousness, the design called for a low engine hood and a larger-than-usual front window. Both features increased the

> **External integrity begins with customers but goes beyond them to include a healthy measure of "market imagination."**

driver's visibility and sense of interaction with the outside world. But the window size also meant that the cabin would get uncomfortably hot on sunny days unless the car had a big air conditioner—as well as a powerful engine to run it.

A large engine—the obvious solution—was precluded by the decision to keep the hood low, since the only suspension system that would work was an expensive, double-wishbone construction that narrowed the engine chamber. And in any case, the engineers wanted the engine to be light so that the car would handle sharply.

The height of the hood became a battlefield, with body, engine, and chassis engineers warring over millimeters. What made the conflict constructive—it ultimately led to the development of a new engine that was both compact and powerful—was the fact that all the combatants understood what the Accord had to achieve. Guided by the large product leader, who saw every argument as an opportunity to reinforce the car's basic concept, the engineers could see their work through future customers' eyes.

As Honda's experience indicates, external integration extends deeply into the development organiza-

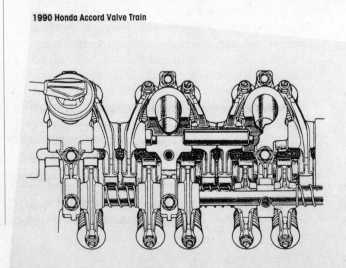

1990 Honda Accord Valve Train

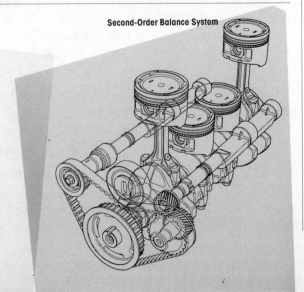

Second-Order Balance System

tion, and it involves much more than being "market oriented" or "customer driven." It begins with customers, to be sure, since the best concept developers invariably supplement the cooked information they get from marketing specialists with raw data they gather themselves. But strong product concepts also include a healthy measure of what we call "market imagination": they encompass what customers say they want and what the concept's creators *imagine* customers will want three or more years into the future. Remembering that customers know only existing products and existing technologies, they avoid the trap of being too close to customers – and designing products that will be out-of-date before they are even manufactured.

Interestingly, companies that are heavily driven by market data tend to slip on external integrity. As a rule, these companies have well-equipped marketing organizations with great expertise in formal research, and they are adept at using data from focus groups, product clinics, and the like to develop customer profiles. But these methods rarely lead to distinctive product concepts. In fact, to the extent that they limit or suppress the imaginations of product designers, they can actually harm a new product's future competitiveness.

How auto companies organize for external integration – and how much power they invest in their integrators – varies greatly. Some companies create an explicit role for an "external integrator" and assign it to people in a few functional units (testers in engineering, for example, and product planners in marketing). Others assign all their external integrators to a single specialized unit, which may be independent or organized by product. Similarly, the work of concept creation and concept realization may be broken up among different groups in the development organization or consolidated under one leader, as it is at Honda.

We have already seen how advantageous consolidating responsibility can be for enhancing external integration. This approach is equally successful in achieving internal integrity.

One of the thorniest issues in creating a strong product concept is when (and how) to involve functional specialists other than those who make up the product development team. As we saw with the Accord, the product concept has clear repercussions for every aspect of the development process, from design and layout to cost and manufacturability. So on the one hand, front-loading input and information from specialists downstream is highly desirable. On the other hand, broad downstream involvement can easily jeopardize the distinctiveness and clarity of a product concept if (as often happens) negotiations

and battles among powerful functions lead to political compromises and patchwork solutions.

The fact that working-level engineers were involved in the concept stage of the Accord's development was essential to its product integrity. Faced with tough choices about the car's front end, the engineers had not only a clear concept to guide them but also one they felt they owned. Moreover, their solution – the new engine – enhanced the Accord's internal integrity by raising its level of technical performance. At the same time, internal demands and functional constraints never compromised the Accord's basic concept. Like many of the other product managers we spoke with, the Accord's product leader knew that democracy without clear concept leadership is the archenemy of distinctive products.

There are other ways to balance downstream expertise with strong concept leadership, of course. (One of Honda's rivals also makes early cross-functional negotiations an important part of its new product development work, for example, but gives a small group of concept creators and assistants six months or so to establish the concept first, before the negotiations begin.) The important point is that integrity depends on striking a balance between the two. Companies that trade off one for the other sacrifice both product and organizational integrity. Those that place sole responsibility for the product concept with a specialized unit (often one within marketing) end up with lots of last-minute design

> Companies that trade off functional expertise for a strong product concept sacrifice integrity.

and engineering changes. Conversely, companies that initiate senior-level, cross-functional negotiations at the very start of every project usually find themselves with undistinguished products.

The integration that leads to product integrity does not surface in organization charts alone, nor is it synonymous with the creation of cross-functional teams, the implementation of "design for manufacturing," or any other useful organizational formula for overhauling development work. Ironically, efforts to increase integration can even undermine it if the integrating mechanisms are misconstructed or if the organization is unprepared for the change. At one U.S. auto company, we found a very coherent cross-functional project team with great spirit and purpose. But the team was made up solely of liaisons and included none of the working engineers actually responsible for drawings and prototypes. So for the most part, engineers ignored the team, whose existence only masked the lack of true integration.

What distinguishes outstanding product developers is the consistency between their formal structures and the informal organization that accomplishes the real work of development. In the case of the Honda Accord, we have seen some important characteristics of such consistency: the company's preference for firsthand information and direct (sometimes conflict-full) discussion; the way specialists are respected but never deified; the constant stream of early, informal communication (even at the risk of creating confusion or inefficiencies in the short run); and, most important, the primacy of strong concept leadership.

Integrity's Champion: The Heavyweight Product Manager

The key to product integrity is leadership. Product managers in companies whose products consistently succeed accomplish two things without fail. They focus the whole development organization on customer satisfaction. And they devise processes (both formal and informal) for creating powerful product concepts and infusing them into the details of production and design. In our lexicon, they are "heavyweight" product managers, and they differ significantly from their lighter weight counterparts in other companies.

During the 1980s, product managers began to appear at more and more of the world's auto companies. In most cases, the title means relatively little. The position adds another box to the organization chart, but the organization's basic structure is still heavily functional. Product managers in these companies coordinate development activities through liaison representatives from each of the engineering departments. They have no direct access to working-level engineers, no contact with marketing, and no concept responsibility. Their positions have less status and power than the functional managers' do, and they have little influence outside of product engineering (and only limited influence within it). Their job is to collect information on the status of work, to help functional groups resolve conflicts, and to facilitate completion of the project's overall goals. They do not actually impair a product's integrity, but neither can they contribute much to it.

The contrast with the heavyweight's job could not be more striking. In a few auto companies, product managers play a role that simply does not ex-

ist in other automakers' development organizations. Like the Accord's large product leader, they are deeply involved in creating a strong product concept. Then, as the concept's guardians, they keep the concept alive and infuse it into every aspect of the new product's design. As one heavyweight product manager told us, "We listen to process engineers. We listen to plant managers. But we make the final decisions. Above all, we cannot make any compromise on the concept. The concept is the soul of the vehicle; we cannot sell it."

Guardianship like this is crucial because the product concept can get lost so easily in the complexity of actually designing, planning, and building a new car. The problems that preoccupied the Accord's product engineers were often almost imperceptibly small: a three-millimeter gap between the window glass and the body; the tiny chips on the car's sills that come from stones kicked up on the road; a minuscule gap between the hood and the body. But problems like these are the stuff of product integrity: all the magic is in the details.

Keeping track of those details, however, is no easy matter. Nor is it easy to keep the product concept fresh and clear in many people's minds during the months (and years) that development consumes. For that reason, heavyweight product managers must be a little like evangelists, with the product concept as their Bible and the work of exhorting, preaching, and reminding as their mission. To paraphrase an assistant product manager in one of the heavyweight organizations, subtle nuances such as the car's taste and character have to be built into the design by fine-tuning. They cannot be expressed completely in planning documents, no matter how detailed those may be. So the product manager has to interact continuously with the engineers to communicate his intentions and to refresh and reinforce their understanding of the product concept.

As concept guardians, heavyweight product managers draw on both personal credibility and expertise and the organizational clout that comes with the job. Themselves engineers by training, heavyweight

> The stuff of product integrity is a minuscule gap between the hood and the car's body— the magic is in the details.

product managers have a broad knowledge of the product and process engineering required to develop an entire vehicle. Years of experience with their companies give their words weight and increase their

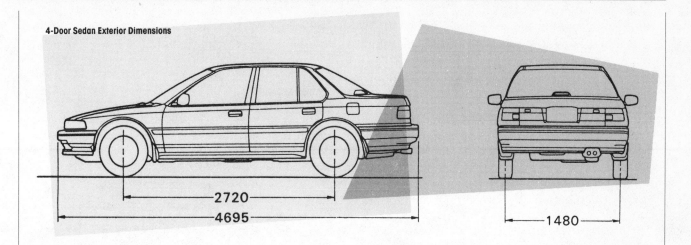

4-Door Sedan Exterior Dimensions

2720
4695
1480

influence with people over whom they have no formal authority.

Product planners and engineers working on the detailed design of specific parts typically fall into this category. Yet as we have seen, the substance of their work is vital to a new car's integrity. To track design decisions and ensure that the concept is being translated accurately, heavyweight managers communicate daily with the functional engineering departments. They also intervene directly when decisions about parts or components that are particularly problematic or central to the product concept are being made. From a functional point of view, this is clearly a breach of organizational etiquette. But in practice, this intervention is usually readily accepted, in part because it is backed by tradition but mostly because of the product manager's credibility. When heavyweights visit bench-level engineers, they come to discuss substantive issues and their input is usually welcome. They are not making courtesy calls or engaging in morale-building exercises.

Organizationally, the heavyweight manager effectively functions as the product's general manager. In addition to concept-related duties, the responsibilities that come with the job include: coordinating production and sales as well as engineering; coordinating the entire project from concept to market; signing off on specification, cost-target, layout, and major component choices; and maintaining direct contact with existing and potential customers. Some of this work occurs through liaison representatives (although the liaisons themselves are "heavier" than they are in the lightweight organizations since they also serve as local project leaders within their functional groups). But there is no mistaking the heavyweights' clout: engineering departments typically report to them (which ones depends on the internal linkages the company wishes to emphasize). Heavyweights are also well supplied with formal

procedures like design review and control of prototype scheduling that give them leverage throughout the organization.

Still, probably the best measure of a product manager's weight is the amount of time that formal meetings and paperwork consume. Lightweight product managers are much like high-level clerks. They spend most of the day reading memos, writing reports, and going to meetings. Heavyweights, in contrast, are invariably "out"—with engineers, plant people, dealers, and customers. "This job can't

> **Heavyweight product managers have to be fluent in the languages of customers, marketers, engineers, and designers.**

be done without wearing out my shoes," one experienced manager commented. "Since I'm asking other engineers for favors, I shouldn't ask them to come to me. I have to go and talk to them."

What lies behind "product managers in motion" is the central role that information plays in bringing new products to life. Take the heavyweight's interaction with customers. Talented product managers spend hours watching people on the street, observing styles, and listening to conversations. Department stores, sports arenas, museums, and discotheques are all part of their "market research" beat.

Heavyweight product managers are equally active in their relations with the test engineers. Like the product manager, test engineers stand in for the customer. When they evaluate a suspension system or test-drive a new car, they are rehearsing the experience the future customer will consume. To do this

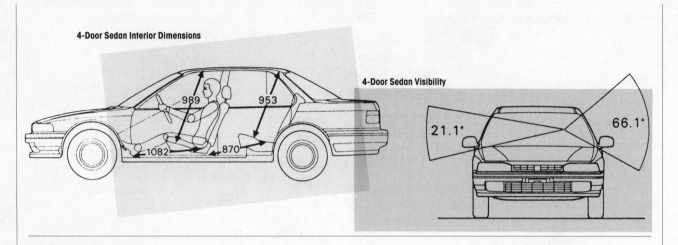

4-Door Sedan Interior Dimensions

989   953

1082   870

4-Door Sedan Visibility

21.1°   66.1°

successfully, in ways that will ensure product integrity, the test engineers must know what to look for. In other words, they must be crystal clear on the product concept.

Heavyweight product managers make sure this clarity exists. They often test-drive vehicles and talk about their experiences with the test engineers. Many can and do evaluate the car's performance on the test track and show up almost daily during critical tests. They also seize every opportunity to build good communication channels and deepen their ties with younger engineers. One product manager said he welcomed disagreements among the test engineers because they gave him a good reason to go out to the proving ground and talk about product concepts with younger people with whom he would not otherwise interact.

If we reverse direction to look at how heavyweight product managers promote internal integrity, the same kind of behavior and activities come to the fore. Direct contact with product engineers and testers, for example, not only reinforces the product concept but also strengthens the links between functions, speeds up decision making and problem solving, and makes it easier to coordinate work flows. In fact, almost everything a product manager does to infuse the concept into the details makes the organization itself work better and faster. The reason is the strong customer orientation that the product concept – and product manager – convey.

The product manager's job touches every part of the new product process. Indeed, heavyweight product managers have to be "multilingual," fluent in the languages of customers, marketers, engineers, and designers. On one side, this means being able to translate an evocative concept like the pocket rocket into specific targets like "maximum speed 250 kilometers per hour" and "drag coefficient less than 0.3" that detail-oriented engineers can easily grasp. On

the other side, it means being able to assess and communicate what a "0.3 drag coefficient" will mean to customers. (The fact that the translation process from customer to engineer is generally harder than that from engineer to customer explains why engineering tends to be the heavyweight product managers' native tongue.)

Because development organizations are continually involved in changing one form of information into another, face-to-face conversations and informal relationships are their life's blood. Heavyweight managers understand this and act on it. Aware that product concepts cannot be communicated in written documents alone (any more than the feel and sensibility of a new car can be captured in words alone), they travel constantly – telling stories, coining phrases, and generally making sure that nothing important gets lost in translation.

The Improvement Ethic

How a company develops new products says a great deal about what that company is and does. For most companies, the journey toward competing on integrity began during the 1980s. Quite possibly, it was inaugurated with a commitment to total quality or to reducing the lead time for developing new products. Heavyweight product management constitutes the next step on that journey. Taking it leads down one of two paths.

Some companies introduce a heavyweight product management system modestly and incrementally. A typical progression might go like this: shift from a strictly functional setup to a lightweight system, with the integrator responsible only for product engineering; expand the product manager's sphere to include new tasks such as product planning or

The Case for Heavyweight Product Management

In the early 1980s, successful products filled the Ford Motor Company's scrapbooks but not its dealers' showrooms. Its cars were widely criticized. Quality was far below competitive standards. Market share was falling. In addition, the company's financial position was woeful, and layoffs were ongoing, among white-collar staff and factory workers alike. By the end of the decade, history was repeating itself: the Ford Explorer, introduced in the spring of 1990, may prove to be Ford's most successful product introduction ever. Despite the fact that it debuted in a down market, the four-door, four-wheel-drive sport-utility vehicle has sold phenomenally well. Rugged yet refined, the Explorer gets all the important details right, from exterior styling to the components and interior design.

Behind the Explorer lay a decade of changes in Ford's management, culture, and product development organization. The changes began in the dark days of the early 1980s with the emergence of new leaders in Ford's executive offices and in design studios. Their herald was the Taurus, introduced in 1985. Designed to be a family vehicle with the styling, handling, and ride of a sophisticated European sedan, the car offered a distinctive yet integrated package in which advanced aerodynamic styling was matched with a newly developed chassis with independent rear suspension and a front-wheel-drive layout. The car's interior, which minimized the chrome and wood paneling that were traditional in American roadsters, had a definite European flavor. So did the ride and the way the car handled: the steering was much more responsive, and the ride was tighter and firmer.

The development efforts that produced the Taurus set in motion profound changes within the Ford engineering, manufacturing, and marketing organizations. Traditionally, Ford's development efforts had been driven by very strong functional managers. In developing the Taurus, however, Ford turned to the "Team Taurus," whose core included principals from all the major functions and activities involved in the creation of the new car. The team was headed by Lew Veraldi, at the time in charge of large-car programs at Ford, and it served to coordinate and integrate the development program at the senior management level.

Team Taurus was the first step on a long path of organizational, attitudinal, and procedural change. As development of the Taurus went ahead, it became clear that integrated development required more than the creation of a team and that there was more to achieving integrity than linking the functions under the direction of a single manager. So the next step in Ford's evolution was the development of the "concept to customer" process, or C to C.

The C to C process took shape during the mid-1980s, as Ford sought aggressively to cut lead time, improve quality, and continue to bring attractive products to market. Led by a handpicked group of engineers and product planners, the C to C project focused on devising a new architecture for product development: its members identified critical milestones, decision points, criteria for decision making, and patterns of responsibility and functional involvement. This architecture was then implemented step by step, in ongoing programs as well as in new efforts.

At about the same time, in 1987, Ford formalized the "program manager" structure that had evolved out of the Taurus experience. (Program manager is the term Ford uses for the position we call product manager.) As part of this structure, senior management affirmed the centrality of cross-functional teams working under the direction of a strong program manager. Moreover, cross-functional integration was reinforced at the operating level as well as at the strategy level. The change in marketing's role is a good example: instead of adding their input through reports and memoranda, marketing people (led by the program manager) meet directly with designers and engineers to discuss concept development and key decisions about features, layout, and components. Similarly, program managers have been given responsibility for critical functions like product planning and layout, where many of the integrative decisions are made.

In successive programs, Ford has refined its approach and pushed integration further and further. The strength of the program managers has also increased. The results are visible in the products Ford developed during the latter part of the 1980s—and in their sales. Beginning with the Taurus, Ford has scored impressive market successes with a number of its new cars: the Lincoln Continental, which expanded Lincoln's share of the luxury market; the Thunderbird Super Coupe, which compares favorably with European high-performance sedans; the Probe, the result of a joint development project with Mazda and which enthusiasts generally rate higher than Mazda's own effort, the MX6; and the sport-utility Explorer.

product-process coordination; then raise the product manager's rank, appoint people with strong reputations to the job, and assign them one project rather than a few to focus their attention and expand their influence. Senior managers that face deep resistance from their functional units often choose this path.

Other companies (particularly smaller players) take a faster, more direct route. One Japanese company leapt to a strong product manager system to introduce a new model. Backed by the widespread belief that the project might well determine the company's future, senior management created an unusually heavy product manager to run it. An executive vice president with many years of experience became the product manager, with department heads from engineering, production, and planning acting as his liaisons and as project leaders within their functional groups. With these changes, management sent a clear signal that the company could no longer survive in its traditional form.

The project succeeded, and today the product is seen as the company's turnaround effort, its reentry as a competitor after years of ineffectual products. The project itself became a model for subsequent changes (including the creation of a product manager office) in the regular development organization.

How a company changes its organization and the speed with which it moves will depend on its position and the competitive threat it faces. But all successful efforts have three common themes: a unifying driver, new blood, and institutional tenacity. (See the insert "The Case for Heavyweight Product Management," which describes Ford Motor Company's progress toward becoming a heavyweight organization.)

Just as engineers need a vision of the overall product to guide their efforts in developing a new car, the people involved in changing an organization need an objective that captures their imaginations. Where changes have taken hold, senior managers have linked them to competition and the drive for tangible advantage in the marketplace.

During the 1980s, the quest for faster development lead time was particularly powerful in driving such efforts. But lead time is not an end in itself. Rather, its pursuit leads people to do things that improve the system overall. In this respect, lead time is like inventory in a just-in-time manufacturing system: reducing work-in-process inventory is somewhat effective, but attacking the root causes of excess inventory truly changes the system.

Companies that successfully focus on lead time generally emphasize changes in internal integration. Product integrity can drive companies to higher performance. Managed well, the drive to create products that fire the imagination gives the implementation of a heavyweight system energy and direction.

Of the many change efforts we have seen, the most successful were led by new people. Some were new to the company, but most came from within

> A company cannot change everyone. It can create new leaders and empower people attuned to new directions.

the organization. Sometimes viewed as mavericks, they saw the potential for change where others saw more of the same. A company cannot change everyone. It can, however, create new leaders and empower people who are attuned to the new direction the company has to take. It can also find nontraditional ways to identify and develop heavyweight product managers for the future, such as apprenticeship systems.

Moving to a heavier product manager structure is a process of discovery—one the U.S. auto company with the ineffectual cross-functional team we described earlier knows very well. Like many others, that company has discovered that changes in organizational structure are important but insufficient. To create a true team, greater change—particularly in the behavior of traditionally powerful functional managers—is needed.

The journey to heavyweight product management is hard, surprisingly so for many managers. Those who succeed do so because they have tenacity. Outstanding companies understand that projects end but the journey doesn't. The challenge to learn from experience and continuously improve is always there.

Yet in company after company, the same problems crop up over and over. Why do most companies learn so little from their product development projects? The explanation is simple: at the end of every project, there is pressure to move on to the next. The cost of this tunnel vision is very high. Those few companies that work at continuous improvement achieve a significant competitive edge. Moving to a more effective development organization can be the basis for instilling an ethic of continuous improvement. Companies that compete on integrity exercise that ethic every day.

Author's note: We gratefully acknowledge the help of Nobuhiko Kawamoto, CEO of Honda Motor Company, and Tateomi Miyoshi, large product leader for the Honda Accord.

David A. Garvin

Competing on the eight dimensions of quality

U.S. managers know that they have to improve the quality of their products because, alas, U.S. consumers have told them so. A survey in 1981 reported that nearly 50% of U.S. consumers believed that the quality of U.S. products had dropped during the previous five years; more recent surveys have found that a quarter of consumers are "not at all" confident that U.S. industry can be depended on to deliver reliable products. Many companies have tried to upgrade their quality, adopting programs that have been staples of the quality movement for a generation: cost of quality calculations, interfunctional teams, reliability engineering, or statistical quality control. Few companies, however, have learned to *compete* on quality. Why?

U.S. consumers doubt that U.S. companies can deliver quality.

Part of the problem, of course, is that until Japanese and European competition intensified, not many companies seriously tried to make quality programs work even as they implemented them. But even if companies *had* implemented the traditional principles of quality control more rigorously, it is doubtful that U.S. consumers would be satisfied today. In my

David A. Garvin is an associate professor of business administration at the Harvard Business School. He has published numerous articles on quality in HBR and other journals and is the recipient of McKinsey Awards for best HBR article in 1982 and 1983. This article draws from his book, Managing Quality, to be published by Free Press.

view, most of those principles were narrow in scope; they were designed as purely defensive measures to preempt failures or eliminate "defects." What managers need now is an aggressive strategy to gain and hold markets, with high quality as a competitive linchpin.

Quality control

To get a better grasp of the defensive character of traditional quality control, we should understand what the quality movement in the United States has achieved so far. How much expense on quality was tolerable? How much "quality" was enough? In 1951, Joseph Juran tackled these questions in the first edition of his *Quality Control Handbook*, a publication that became the quality movement's bible. Juran observed that quality could be understood in terms of avoidable and unavoidable costs: the former resulted from defects and product failures like scrapped materials or labor hours required for rework, repair, and complaint processing; the latter were associated with prevention, i.e., inspection, sampling, sorting, and other quality control initiatives. Juran regarded failure costs as "gold in the mine" because they could be reduced sharply by investing in quality improvement. He estimated that avoidable quality losses typically ranged from $500 to $1,000 per productive operator per year—big money back in the 1950s.

Reading Juran's book, executives inferred roughly how much to invest in quality improvement: expenditures on prevention were justified if they were lower than the costs of product failure. A corollary principle was that decisions made early in the production chain (e.g., when engineers first sketched out a product's design) have implications for the level of

"I spoke to my attorney today, Wendell, and I'm thinking of putting you into play."

quality costs incurred later, both in the factory and the field.

In 1956, Armand Feigenbaum took Juran's ideas a step further by proposing "total quality control" (TQC). Companies would never make high-quality products, he argued, if the manufacturing department were forced to pursue quality in isolation. TQC called for "interfunctional teams" from marketing, engineering, purchasing, and manufacturing. These teams would share responsibility for all phases of design and manufacturing and would disband only when they had placed a product in the hands of a satisfied customer—who remained satisfied.

Feigenbaum noted that all new products moved through three stages of activity: design control, incoming material control, and product or shop-floor control. This was a step in the right direction. But Feigenbaum did not really consider how quality was first of all a strategic question for any business; how, for instance, quality might govern the development of a design and the choice of features or options. Rather, design control meant for Feigenbaum mainly preproduction assessments of a new design's manufacturability, or that projected manufacturing techniques should be debugged through pilot runs. Materials control included vendor evaluations and incoming inspection procedures.

In TQC, quality was a kind of burden to be shared—no single department shouldered all the responsibility. Top management was ultimately accountable for the effectiveness of the system; Feigenbaum, like Juran, proposed careful reporting of the costs of quality to senior executives in order to ensure their commitment. The two also stressed statistical approaches to quality, including process control charts that set limits to acceptable variations in key variables affecting a product's production. They endorsed sampling procedures that allowed managers to draw inferences about the quality of entire batches of products from the condition of items in a small, randomly selected sample.

Despite their attention to these techniques, Juran, Feigenbaum, and other experts like W. Edwards Deming were trying to get managers to see beyond purely statistical controls on quality. Meanwhile, another branch of the quality movement emerged, relying even more heavily on probability theory and statistics. This was "reliability engineering," which originated in the aerospace and electronics industries.

In 1950, only one-third of the U.S. Navy's electronic devices worked properly. A subsequent study by the Rand Corporation estimated that every vacuum tube the military used had to be backed by nine others in warehouses or on order. Reliability engineering addressed these problems by adapting the laws of probability to the challenge of predicting equipment stress.

Reliability engineering measures led to:

- Techniques for reducing failure rates while products were still in the design stage.

- Failure mode and effect analysis, which systematically reviewed how alternative designs could fail.

- Individual component analysis, which computed the failure probability of key components and aimed to eliminate or strengthen the weakest links.

- Derating, which required that parts be used below their specified stress levels.

- Redundancy, which called for a parallel system to back up an important component or subsystem in case it failed.

Naturally, an effective reliability program required managers to monitor field failures closely to give company engineers the information needed to plan new designs. Effective field failure reporting also demanded the development of systems of data collection, including return of failed parts to the laboratory for testing and analysis.

Now, the proponents of all these approaches to quality control might well have denied that their views of quality were purely defensive. But what else was implied by the solutions they stressed – material controls, outgoing batch inspections, stress tests? Perhaps the best way to see the implications of their logic is in traditional quality control's most extreme form, a program called "Zero Defects." No other program defined quality so stringently as an absence of failures – and no wonder, since it emerged from the defense industries where the product was a missile whose flawless operation was, for obvious reasons, imperative.

In 1961, the Martin Company was building Pershing missiles for the U.S. Army. The design of the missile was sound, but Martin found that it could maintain high quality only through a massive program of inspection. It decided to offer workers incentives to lower the defect rate, and in December 1961, delivered a Pershing missile to Cape Canaveral with "zero discrepancies." Buoyed by this success, Martin's general manager in Orlando, Florida accepted a challenge, issued by the U.S. Army's missile command, to deliver the first field Pershing one month ahead of schedule. But he went even further. He promised that the missile would be perfect, with no hardware problems or document errors, and that all equipment would be fully operational 10 days after delivery (the norm was 90 days or more).

> **Quality means
> pleasing consumers,
> not just protecting them
> from annoyances.**

Two months of feverish activity followed; Martin asked all employees to contribute to building the missile exactly right the first time since there would be virtually no time for the usual inspections. Management worked hard to maintain enthusiasm on the plant floor. In February 1962, Martin delivered on time a perfect missile that was fully operational in less than 24 hours.

This experience was eye-opening for both Martin and the rest of the aerospace industry. After careful review, management concluded that, in effect, its own changed attitude had assured the project's success. In the words of one close observer: "The one time management demanded perfection, it happened!"[1]

Martin management thereafter told employees that the only acceptable quality standard was "zero defects." It instilled this principle in the work force through training, special events, and by posting quality results. It set goals for workers and put great effort into giving each worker positive criticism. Formal techniques for problem solving, however, remained limited. For the most part, the program focused on motivation – on changing the attitudes of employees.

Strategic quality management

On the whole, U.S. corporations did not keep pace with quality control innovations the way a number of overseas competitors did. Particularly after World War II, U.S. corporations expanded rapidly and many became complacent. Managers knew that consumers wouldn't drive a VW Beetle, indestructible as it was, if they could afford a fancier car – even if this meant more visits to the repair shop.

But if U.S. car manufacturers *had* gotten their products to outlast Beetles, U.S. quality managers still would not have been prepared for Toyota Corollas – or Sony televisions. Indeed, there was nothing in the principles of quality control to disabuse them of the idea that quality was merely something that could hurt a company if ignored; that added quality was the designer's business – a matter, perhaps, of chrome and push buttons.

The beginnings of strategic quality management cannot be dated precisely because no single book or article marks its inception. But even more than in consumer electronics and cars, the volatile market in semiconductors provides a telling example of change. In March 1980, Richard W. Anderson, general manager of Hewlett-Packard's Data Systems Division, reported that after testing 300,000 16K RAM chips from three U.S. and three Japanese manufacturers, Hewlett-Packard had discovered wide disparities in quality. At incoming inspection, the Japanese chips had a failure rate of zero; the comparable rate for the three U.S. manufacturers was between 11 and 19 failures per 1,000. After 1,000 hours of use, the failure rate of the Japanese chips was between 1 and 2 per 1,000; usable U.S. chips failed up to 27 times per thousand.

Several U.S. semiconductor companies reacted to the news impulsively, complaining that the Japanese were sending only their best components to

---

1 James F. Halpin,
Zero Defects
(New York:
McGraw-Hill, 1966), p. 15.

the all-important U.S. market. Others disputed the basic data. The most perceptive market analysts, however, noted how differences in quality coincided with the rapid ascendancy of Japanese chip manufacturers. In a few years the Japanese had gone from a standing start to significant market shares in both the 16K and 64K chip markets. Their message—intentional or not—was that quality could be a potent strategic weapon.

U.S. semiconductor manufacturers got the message. In 16K chips the quality gap soon closed. And in industries as diverse as machine tools and radial tires, each of which had seen its position erode in the face of Japanese competition, there has been a new seriousness about quality too. But how to translate seriousness into action? Managers who are now determined to compete on quality have been thrown back on the old questions: How much quality is enough? What does it take to look at quality from the customer's vantage point? These are still hard questions today.

Some consumer preferences should be treated as absolute performance standards.

To achieve quality gains, I believe, managers need a new way of thinking, a conceptual bridge to the consumer's vantage point. Obviously, market studies acquire a new importance in this context, as does a careful review of competitors' products. One thing is certain: high quality means pleasing consumers, not just protecting them from annoyances. Product designers, in turn, should shift their attention from prices at the time of purchase to life cycle costs that include expenditures on service and maintenance—the customer's total costs. Even consumer complaints play a new role because they provide a valuable source of product information.

But managers have to take a more preliminary step—a crucial one, however obvious it may appear. They must first develop a clear vocabulary with which to discuss quality as *strategy*. They must break down the word quality into manageable parts. Only then can they define the quality niches in which to compete.

I propose eight critical dimensions or categories of quality that can serve as a framework for strategic analysis: performance, features, reliability, conformance, durability, serviceability, aesthetics, and perceived quality.[2] Some of these are always mutually reinforcing; some are not. A product or service can rank high on one dimension of quality and low on another—indeed, an improvement in one may be achieved only at the expense of another. It is precisely this interplay that makes strategic quality management possible; the challenge to managers is to compete on selected dimensions.

---

1   Performance

Of course, performance refers to a product's primary operating characteristics. For an automobile, performance would include traits like acceleration, handling, cruising speed, and comfort; for a television set, performance means sound and picture clarity, color, and the ability to receive distant stations. In service businesses—say, fast food and airlines—performance often means prompt service.

Because this dimension of quality involves measurable attributes, brands can usually be ranked objectively on individual aspects of performance. Overall performance rankings, however, are more difficult to develop, especially when they involve benefits that not every consumer needs. A power shovel with a capacity of 100 cubic yards per hour will "outperform" one with a capacity of 10 cubic yards per hour. Suppose, however, that the two shovels possessed the identical capacity—60 cubic yards per hour—but achieved it differently: one with a 1-cubic-yard bucket operating at 60 cycles per hour, the other with a 2-cubic-yard bucket operating at 30 cycles per hour. The capacities of the shovels would then be the same, but the shovel with the larger bucket could handle massive boulders while the shovel with the smaller bucket could perform precision work. The "superior performer" depends entirely on the task.

Some cosmetics wearers judge quality by a product's resistance to smudging; others, with more sensitive skin, assess it by how well it leaves skin irritation-free. A 100-watt light bulb provides greater candlepower than a 60-watt bulb, yet few customers would regard the difference as a measure of quality. The bulbs simply belong to different performance classes. So the question of whether performance differences are quality differences may depend on circumstantial preferences—but preferences based on functional requirements, not taste.

Some performance standards *are* based on subjective preferences, but the preferences are so universal that they have the force of an objective standard. The quietness of an automobile's ride is usually viewed as a direct reflection of its quality. Some people like a dimmer room, but who wants a noisy car?

---

2 This framework first appeared,
  in a preliminary form,
  in my article
  "What Does 'Product Quality' Really Mean?"
  Sloan Management Review, Fall 1984.

2   Features

Similar thinking can be applied to features, a second dimension of quality that is often a secondary aspect of performance. Features are the "bells and whistles" of products and services, those characteristics that supplement their basic functioning. Examples include free drinks on a plane, permanent-press cycles on a washing machine, and automatic tuners on a color television set. The line separating primary performance characteristics from secondary features is often difficult to draw. What is crucial, again, is that features involve objective and measurable attributes; objective individual needs, not prejudices, affect their translation into quality differences.

To many customers, of course, superior quality is less a reflection of the availability of particular features than of the total number of options available. Often, choice is quality: buyers may wish to customize or personalize their purchases. Fidelity Investments and other mutual fund operators have pursued this more "flexible" approach. By offering their clients a wide range of funds covering such diverse fields as health care, technology, and energy – and by then encouraging clients to shift savings among these – they have virtually tailored investment portfolios.

Employing the latest in flexible manufacturing technology, Allen-Bradley customizes starter motors for its buyers without having to price its products prohibitively. Fine furniture stores offer their customers countless variations in fabric and color. Such strategies impose heavy demands on operating managers; they are an aspect of quality likely to grow in importance with the perfection of flexible manufacturing technology.

3   Reliability

This dimension reflects the probability of a product malfunctioning or failing within a specified time period. Among the most common measures of reliability are the mean time to first failure, the mean time between failures, and the failure rate per unit time. Because these measures require a product to be in use for a specified period, they are more relevant to durable goods than to products and services that are consumed instantly.

Reliability normally becomes more important to consumers as downtime and maintenance become more expensive. Farmers, for example, are especially sensitive to downtime during the short harvest season. Reliable equipment can mean the difference between a good year and spoiled crops. But consumers in other markets are more attuned than ever to product reliability too. Computers and copying machines certainly compete on this basis. And recent market research shows that, especially for young women, reliability has become an automobile's most desired attribute. Nor is the government, our biggest single consumer, immune. After seeing its expenditures for major weapons repair jump from $7.4 billion in fiscal year 1980 to $14.9 billion in fiscal year 1985, the Department of Defense has begun cracking down on contractors whose weapons fail frequently in the field.

4   Conformance

A related dimension of quality is conformance, or the degree to which a product's design and operating characteristics meet established standards. This dimension owes the most to the traditional approaches to quality pioneered by experts like Juran.

All products and services involve specifications of some sort. When new designs or models are developed, dimensions are set for parts and purity standards for materials. These specifications are normally expressed as a target or "center"; deviance from the center is permitted within a specified range. Because this approach to conformance equates good quality with operating inside a tolerance band, there is little interest in whether specifications have been met exactly. For the most part, dispersion within specification limits is ignored.

One drawback of this approach is the problem of "tolerance stack-up": when two or more parts are to be fit together, the size of their tolerances often determines how well they will match. Should one part fall at a lower limit of its specification, and a matching part at its upper limit, a tight fit is unlikely. Even if the parts are rated acceptable initially, the link between them is likely to wear more quickly than one made from parts whose dimensions have been centered more exactly.

To address this problem, a more imaginative approach to conformance has emerged. It is closely associated with Japanese manufacturers and the work of Genichi Taguchi, a prizewinning Japanese statistician. Taguchi begins with the idea of "loss function," a measure of losses from the time a product is shipped. (These losses include warranty costs, nonrepeating customers, and other problems resulting from performance failure.) Taguchi then compares such losses to two alternative approaches to quality: on the one hand, simple conformance to specifications, and on the other, a measure of the degree to which parts or products diverge from the ideal target or center.

He demonstrates that "tolerance stack-up" will be worse – more costly – when the dimensions

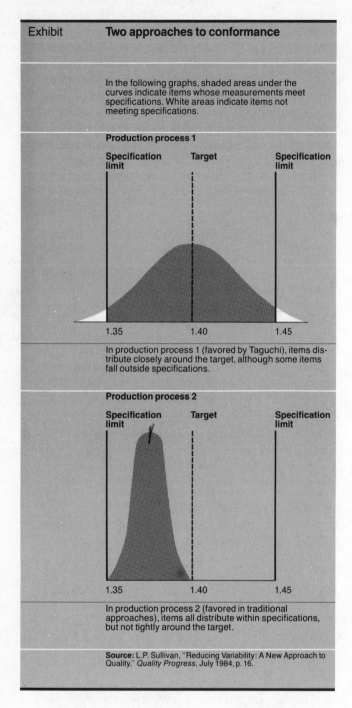

Exhibit   **Two approaches to conformance**

In the following graphs, shaded areas under the curves indicate items whose measurements meet specifications. White areas indicate items not meeting specifications.

Production process 1

Specification limit | Target | Specification limit

1.35   1.40   1.45

In production process 1 (favored by Taguchi), items distribute closely around the target, although some items fall outside specifications.

Production process 2

Specification limit | Target | Specification limit

1.35   1.40   1.45

In production process 2 (favored in traditional approaches), items all distribute within specifications, but not tightly around the target.

Source: L.P. Sullivan, "Reducing Variability: A New Approach to Quality," *Quality Progress*, July 1984, p. 16.

of parts are more distant from the center than when they cluster around it, even if some parts fall outside the tolerance band entirely. According to Taguchi's approach, production process 1 in the *Exhibit* is better even though some items fall beyond specification limits. Traditional approaches favor production process 2. The challenge for quality managers is obvious.

Incidentally, the two most common measures of failure in conformance—for Taguchi and everyone else—are defect rates in the factory and, once a product is in the hands of the customer, the incidence of service calls. But these measures neglect other devi-ations from standard, like misspelled labels or shoddy construction, that do not lead to service or repair. In service businesses, measures of conformance normally focus on accuracy and timeliness and include counts of processing errors, unanticipated delays, and other frequent mistakes.

---

5   Durability

A measure of product life, durability has both economic and technical dimensions. Technically, durability can be defined as the amount of use one gets from a product before it deteriorates. After so many hours of use, the filament of a light bulb burns up and the bulb must be replaced. Repair is impossible. Economists call such products "one-hoss shays" (after the carriage in the Oliver Wendell Holmes poem that was designed by the deacon to last a hundred years, and whose parts broke down simultaneously at the end of the century).

In other cases, consumers must weigh the expected cost, in both dollars and personal inconvenience, of future repairs against the investment and operating expenses of a newer, more reliable model. Durability, then, may be defined as the amount of use one gets from a product before it breaks down and replacement is preferable to continued repair.

This approach to durability has two important implications. First, it suggests that durability and reliability are closely linked. A product that often fails is likely to be scrapped earlier than one that is more reliable; repair costs will be correspondingly higher and the purchase of a competitive brand will look that much more desirable. Because of this linkage, companies sometimes try to reassure customers by offering lifetime guarantees on their products, as 3M has done with its videocassettes. Second, this approach implies that durability figures should be interpreted with care. An increase in product life may not be the result of technical improvements or the use of longer-lived materials. Rather, the underlying economic environment simply may have changed.

For example, the expected life of an automobile rose during the last decade—it now averages 14 years—mainly because rising gasoline prices and a weak economy reduced the average number of miles driven per year. Still, durability varies widely among brands. In 1981, estimated product lives for major home appliances ranged from 9.9 years (Westinghouse) to 13.2 years (Frigidaire) for refrigerators, 5.8 years (Gibson) to 18 years (Maytag) for clothes washers, 6.6 years (Montgomery Ward) to 13.5 years (Maytag) for dryers, and 6 years (Sears) to 17 years (Kirby) for vacuum cleaners.[3] This wide dispersion suggests that durability is a potentially fertile area for further quality differentiation.

6 Serviceability

A sixth dimension of quality is serviceability, or the speed, courtesy, competence, and ease of repair. Consumers are concerned not only about a product breaking down but also about the time before service is restored, the timeliness with which service appointments are kept, the nature of dealings with service personnel, and the frequency with which service calls or repairs fail to correct outstanding problems. In those cases where problems are not immediately resolved and complaints are filed, a company's complaint-handling procedures are also likely to affect customers' ultimate evaluation of product and service quality.

Some of these variables reflect differing personal standards of acceptable service. Others can be measured quite objectively. Responsiveness is typically measured by the mean time to repair, while technical competence is reflected in the incidence of multiple service calls required to correct a particular problem. Because most consumers equate rapid repair and reduced downtime with higher quality, these elements of serviceability are less subject to personal interpretation than are those involving evaluations of courtesy or standards of professional behavior.

Even reactions to downtime, however, can be quite complex. In certain environments, rapid response becomes critical only after certain thresholds have been reached. During harvest season, farmers generally accept downtime of one to six hours on harvesting equipment, such as combines, with little resistance. As downtime increases, they become anxious; beyond eight hours of downtime they become frantic and frequently go to great lengths to continue harvesting even if it means purchasing or leasing additional equipment. In markets like this, superior service can be a powerful selling tool. Caterpillar guarantees delivery of repair parts anywhere in the world within 48 hours; a competitor offers the free loan of farm equipment during critical periods should its customers' machines break down.

Customers may remain dissatisfied even after completion of repairs. How these complaints are handled is important to a company's reputation for quality and service. Eventually, profitability is likely to be affected as well. A 1976 consumer survey found that among households that initiated complaints to resolve problems, more than 40% were not satisfied with the results. Understandably, the degree of satisfaction with complaint resolution closely correlated with consumers' willingness to repurchase the offending brands.[4]

Companies differ widely in their approaches to complaint handling and in the importance they attach to this element of serviceability. Some do their best to resolve complaints; others use legal gimmicks, the silent treatment, and similar ploys to rebuff dissatisfied customers. Recently, General Electric, Pillsbury, Procter & Gamble, Polaroid, Whirlpool, Johnson & Johnson, and other companies have sought to preempt consumer dissatisfaction by installing toll-free telephone hot lines to their customer relations departments.

7 Aesthetics

The final two dimensions of quality are the most subjective. Aesthetics—how a product looks, feels, sounds, tastes, or smells—is clearly a matter of personal judgment and a reflection of individual preference. Nevertheless, there appear to be some patterns in consumers' rankings of products on the basis of taste. A recent study of quality in 33 food categories, for example, found that high quality was most often associated with "rich and full flavor, tastes natural, tastes fresh, good aroma, and looks appetizing."[5]

The aesthetics dimension differs from subjective criteria pertaining to "performance"—the quiet car engine, say—in that aesthetic choices are not nearly universal. Not all people prefer "rich and full" flavor or even agree on what it means. Companies therefore have to search for a niche. On this dimension of quality, it is impossible to please everyone.

8 Perceived quality

Consumers do not always have complete information about a product's or service's attributes; indirect measures may be their only basis for comparing brands. A product's durability, for example, can seldom be observed directly; it usually must be inferred from various tangible and intangible aspects of the product. In such circumstances, images, advertising, and brand names—inferences about quality rather than the reality itself—can be critical. For this reason, both Honda—which makes cars in Marysville, Ohio—and Sony—which builds color televisions in San Diego—have been reluctant to publicize that their products are "made in America."

Reputation is the primary stuff of perceived quality. Its power comes from an unstated anal-

3 Roger B. Yepsen, Jr., ed.,
The Durability Factor
(Emmaus, Penn:
Rodale Press, 1982), p. 190.

4 TARP, *Consumer Complaint
Handling in America: Final Report*
(Springfield, Va.:
National Technical Information Service,
U.S. Department of Commerce, 1979).

5 P. Greg Bonner and Richard Nelson,
"Product Attributes and
Perceived Quality: Foods," in
Perceived Quality,
ed. Jacob Jacoby and Jerry C. Olson
(Lexington, Mass.:
Lexington Books, D.C. Heath, 1985), p. 71.

ogy: that the quality of products today is similar to the quality of products yesterday, or the quality of goods in a new product line is similar to the quality of a company's established products. In the early 1980s, Maytag introduced a new line of dishwashers. Needless to say, salespeople immediately emphasized the product's reliability—not yet proven—because of the reputation of Maytag's clothes washers and dryers.

Competing on quality

This completes the list of the eight dimensions of quality. The most traditional notions—conformance and reliability—remain important, but they are subsumed within a broader strategic framework. A company's first challenge is to use this framework to explore the opportunities it has to distinguish its products from another company's wares.

The quality of an automobile tire may reflect its tread-wear rate, handling, traction in dangerous driving conditions, rolling resistance (i.e., impact on gas mileage), noise levels, resistance to punctures, or appearance. High-quality furniture may be distinguished by its uniform finish, an absence of surface flaws, reinforced frames, comfort, or superior design.

One company's quality niche may be another's trap.

Even the quality of a less tangible product like computer software can be evaluated in multiple dimensions. These dimensions include reliability, ease of maintenance, match with users' needs, integrity (the extent to which unauthorized access can be controlled), and portability (the ease with which a program can be transferred from one hardware or software environment to another).

A company need not pursue all eight dimensions simultaneously. In fact, that is seldom possible unless it intends to charge unreasonably high prices. Technological limitations may impose a further constraint. In some cases, a product or service can be improved in one dimension of quality only if it becomes worse in another. Cray Research, a manufacturer of supercomputers, has faced particularly difficult choices of this sort. According to the company's chairman, if a supercomputer doesn't fail every month or so, it probably wasn't built for maximum speed;

in pursuit of higher speed, Cray has deliberately sacrificed reliability.

There are other trade-offs. Consider the following:

☐ In entering U.S. markets, Japanese manufacturers often emphasize their products' reliability and conformance while downplaying options and features. The superior "fits and finishes" and low repair rates of Japanese cars are well known; less often recognized are their poor safety records and low resistance to corrosion.

☐ Tandem Computers has based its business on superior reliability. For computer users that find downtime intolerable, like telephone companies and utilities, Tandem has devised a fail-safe system: two processors working in parallel and linked by software that shifts responsibility between the two if an important component or subsystem fails. The result, in an industry already well-known for quality products, has been spectacular corporate growth. In 1984, after less than 10 years in business, Tandem's annual sales topped $500 million.

☐ Not long ago, New York's Chemical Bank upgraded its services for collecting payments for corporations. Managers had first conducted a user survey indicating that what customers wanted most was rapid response to queries about account status. After it installed a computerized system to answer customers' calls, Chemical, which banking consumers had ranked fourth in quality in the industry, jumped to first.

☐ In the piano business, Steinway & Sons has long been the quality leader. Its instruments are known for their even voicing (the evenness of character and timbre in each of the 88 notes on the keyboard), the sweetness of their registers, the duration of their tone, their long lives, and even their fine cabinet work. Each piano is built by hand and is distinctive in sound and style. Despite these advantages, Steinway recently has been challenged by Yamaha, a Japanese manufacturer that has built a strong reputation for quality in a relatively short time. Yamaha has done so by emphasizing reliability and conformance, two quality dimensions that are low on Steinway's list.

These examples confirm that companies can pursue a selective quality niche. In fact, they may have no other choice, especially if competitors have established reputations for a certain kind of excellence. Few products rank high on all eight dimensions of quality. Those that do—Cross pens, Rolex watches, Rolls-Royce automobiles—require consumers to pay the cost of skilled workmanship.

6 Consumer Network, Inc., *Brand Quality Perceptions* (Philadelphia: Consumer Network, August 1983), p. 17 and 50-51.

Strategic errors

A final word, not about strategic opportunities, but about the worst strategic mistakes. The first is direct confrontation with an industry's leader. As with Yamaha vs. Steinway, it is far preferable to nullify the leader's advantage in a particular niche while avoiding the risk of retaliation. Moreover, a common error is to introduce dimensions of quality that are unimportant to consumers. When deregulation unlocked the market for residential telephones, a number of manufacturers, including AT&T, assumed that customers equated quality with a wide range of expensive features. They were soon proven wrong. Fancy telephones sold poorly while durable, reliable, and easy-to-operate sets gained large market shares.

Shoddy market research can add quality features nobody wants.

Shoddy market research often results in neglect of quality dimensions that *are* critical to consumers. Using outdated surveys, car companies overlooked how important reliability and conformance were becoming in the 1970s; ironically, these companies failed consumers on the very dimensions that were key targets of traditional approaches to quality control.

It is often a mistake to stick with old quality measures when the external environment has changed. A major telecommunications company had always evaluated its quality by measuring timeliness — the amount of time it took to provide a dial tone, to connect a call, or to be connected to an operator. On these measures it performed well. More sophisticated market surveys, conducted in anticipation of the industry's deregulation, found that consumers were not really concerned about call connection time; consumers assumed that this would be more or less acceptable. They were more concerned with the clarity of transmission and the degree of static on the line. On these measures, the company found it was well behind its competitors.

In an industry like semiconductor manufacturing equipment, Japanese machines generally require less set-up time; they break down less often and have few problems meeting their specified performance levels. These are precisely the traits desired by most buyers. Still, U.S. equipment can *do* more. As one U.S. plant manager put it: "Our equipment is more advanced, but Japanese equipment is more developed."

Quality measures may be inadequate in less obvious ways. Some measures are too limited; they fail to capture aspects of quality that are important for competitive success. Singapore International Airlines, a carrier with a reputation for excellent service, saw its market share decline in the early 1980s. The company dismissed quality problems as the cause of its difficulties because data on service complaints showed steady improvement during the period. Only later, after SIA solicited consumer responses, did managers see the weakness of their former measures. Relative declines in service had indeed been responsible for the loss of market share. Complaint counts had failed to register problems because the proportion of passengers who wrote complaint letters was small — they were primarily Europeans and U.S. citizens rather than Asians, the largest percentage of SIA passengers. SIA also had failed to capture data about its competitors' service improvements.

The pervasiveness of these errors is difficult to determine. Anecdotal evidence suggests that many U.S. companies lack hard data and are thus more vulnerable than they need be. One survey found that 65% of executives thought that consumers could readily name — without help — a good quality brand in a big-ticket category like major home appliances. But when the question was actually posed to consumers, only 16% could name a brand for small appliances, and only 23% for large appliances.[6] Are U.S. executives that ill-informed about consumers' perceptions? The answer is not likely to be reassuring.

Managers have to stop thinking about quality merely as a narrow effort to gain control of the production process, and start thinking more rigorously about consumers' needs and preferences. Quality is not simply a problem to be solved; it is a competitive opportunity. ▱

Reprint 87603

Design products not to fail in the field; you will simultaneously reduce defectives in the factory.

Robust Quality

by Genichi Taguchi and Don Clausing

When a product fails, you must replace it or fix it. In either case, you must track it, transport it, and apologize for it. Losses will be much greater than the costs of manufacture, and none of this expense will necessarily recoup the loss to your reputation. Taiichi Ohno, the renowned former executive vice president of Toyota Motor Corporation, put it this way: Whatever an executive thinks the losses of poor quality are, they are actually six times greater.

How can manufacturing companies minimize them? If U.S. managers learn only one new principle from the collection now known as Taguchi Methods, let it be this: Quality is a virtue of design. The "robustness" of products is more a function of good

In 1989, Genichi Taguchi received MITI's Purple Ribbon Award from the emperor of Japan for his contribution to Japanese industrial standards. He is executive director of the American Supplier Institute and director of the Japan Industrial Technology Transfer Association. He is also the author of The System of Experimental Design *(ASI Press, 1987). Don Clausing, formerly of Xerox, is the Bernard M. Gordon Adjunct Professor of Engineering Innovation and Practice at MIT. He has edited Mr. Taguchi's works in English and is a leading exponent of his views. He is the author (with John R. Hauser) of "The House of Quality" (HBR May-June 1988).*

design than of on-line control, however stringent, of manufacturing processes. Indeed—though not nearly so obvious—an inherent lack of robustness in product design is the primary driver of superfluous manufacturing expenses. But managers will have to learn more than one principle to understand why.

Zero Defects, Imperfect Products

For a generation, U.S. managers and engineers have reckoned quality losses as equivalent to the costs absorbed by the factory when it builds defective products—the squandered value of products that cannot be shipped, the added costs of rework, and so on. Most managers think losses are low when the factory ships pretty much what it builds; such is the message of statistical quality control and other traditional quality control programs that we'll subsume under the seductive term "Zero Defects."

Of course, customers do not give a hang about a factory's record of staying "in spec" or minimizing scrap. For customers, the proof of a product's quality is in its performance when rapped, overloaded, dropped, and splashed. Then, too many products dis-

Taguchi's Quality Imperatives

☐ Quality losses result mainly from product failure after sale; product "robustness" is more a function of product design than of on-line control, however stringent, of manufacturing processes.

☐ Robust products deliver a strong "signal" regardless of external "noise" and with a minimum of internal "noise." Any strengthening of a design, that is, any marked increase in the signal-to-noise ratios of component parts, will simultaneously improve the robustness of the product as a whole.

☐ To set targets at maximum signal-to-noise ratios, develop a system of trials that allows you to analyze change in overall system performance according to the *average* effect of change in component parts, that is, when you subject parts to varying values, stresses, and experimental conditions. In new products, average effects may be most efficiently discerned by means of "orthogonal arrays."

☐ To build robust products, set ideal target values for components and then minimize the average of the square of deviations for combined components, averaged over the various customer-use conditions.

☐ Before products go on to manufacturing, tolerances are set. Overall quality loss then increases by the square of deviation from the target value, that is, by the quadratic formula $L = D^2C$, where the constant, C, is determined by the cost of the countermeasure that might be employed in the factory. This is the "Quality Loss Function."

☐ You gain virtually nothing in shipping a product that just barely satisfies the corporate standard over a product that just fails. Get on target, don't just try to stay in-spec.

☐ Work relentlessly to achieve designs that can be produced consistently; demand consistency from the factory. Catastrophic stack-up is more likely from scattered deviation within specifications than from consistent deviation outside. Where deviation from target is consistent, adjustment to the target is possible.

☐ A concerted effort to reduce product failure in the field will simultaneously reduce the number of defectives in the factory. Strive to reduce variances in the components of the product and you will reduce variances in the production system as a whole.

☐ Competing proposals for capital equipment or competing proposals for on-line interventions may be compared by adding the cost of each proposal to the average quality loss, that is, the deviations expected from it.

play temperamental behavior and annoying or even dangerous performance degradations. We all prefer copiers whose copies are clear under low power; we all prefer cars designed to steer safely and predictably, even on roads that are wet or bumpy, in crosswinds, or with tires that are slightly under or overinflated. We say these products are robust. They gain steadfast customer loyalty.

Design engineers take for granted environmental forces degrading performance (rain, low voltage, and the like). They try to counter these effects in product design—insulating wires, adjusting tire treads, sealing joints. But some performance degradations come from the interaction of parts themselves, not from anything external happening to them. In an ideal product—in an ideal anything—parts work in perfect

Ambient variations in the factory are rarely as damaging as variations in customer use.

harmony. Most real products, unfortunately, contain perturbations of one kind or another, usually the result of a faulty meshing of one component with corresponding components. A drive shaft vibrates and wears out a universal joint prematurely; a fan's motor generates too much heat for a sensitive microprocessor.

Such performance degradations may result either from something going wrong in the factory or from an inherent failure in the design. A drive shaft may vibrate too much because of a misaligned lathe or a misconceived shape; a motor may prove too hot because it was put together improperly or yanked into the design impetuously. Another way of saying this is that work-in-progress may be subjected to wide variations in factory process and ambience, and products may be subjected to wide variations in the conditions of customer use.

Why do we insist that most degradations result from the latter kind of failure, design failures, and not from variations in the factory? Because the ambient or process variations that work-in-process may be subjected to in the factory are not nearly as dramatic as the variations that products are subjected to in a customer's hands—obvious when you think about it, but how many exponents of Zero Defects do? Zero Defects says, The effort to reduce process failure in the factory will simultaneously reduce instances of product failure in the field. We say, The effort to reduce product failure in the field will simultaneously reduce the number of defectives in the factory.

Still, we can learn something interesting about the roots of robustness and the failures of traditional quality control by confronting Zero Defects on its own ground. It is in opposition to Zero Defects that Taguchi Methods emerged.

Robustness as Consistency

According to Zero Defects, designs are essentially fixed before the quality program makes itself felt; serious performance degradations result from the failure of parts to mate and interface just so. When manufacturing processes are out of control, that is, when there are serious variations in the manufacture of parts, products cannot be expected to perform well in the field. Faulty parts make faulty connections. A whole product is the sum of its connections.

Of course, no two drive shafts can be made *perfectly* alike. Engineers working within the logic of Zero Defects presuppose a certain amount of variance in the production of any part. They specify a target for a part's size and dimension, then tolerances that they presume will allow for trivial deviations from this target. What's wrong with a drive shaft that should be 10 centimeters in diameter actually coming in at 9.998?

Nothing. The problem – and it is widespread – comes when managers of Zero Defects programs make a virtue of this necessity. They grow accustomed to thinking about product quality in terms of acceptable deviation from targets – instead of the consistent effort to hit them. Worse, managers may specify tolerances that are much too wide because they assume it would cost too much for the factory to narrow them.

Consider the case of Ford vs. Mazda (then known as Toyo Koygo), which unfolded just a few years ago. Ford owns about 25% of Mazda and asked the Japanese company to build transmissions for a car it was selling in the United States. Both Ford and Mazda were supposed to build to identical specifications; Ford adopted Zero Defects as its standard. Yet after the cars had been on the road for a while, it became clear that Ford's transmissions were generating far higher warranty costs and many more customer complaints about noise.

To its credit, Ford disassembled and carefully measured samples of transmissions made by both companies. At first, Ford engineers thought their gauges were malfunctioning. Ford parts were all in-spec, but Mazda gearboxes betrayed no variability at all from targets. Could *that* be why Mazda incurred lower production, scrap, rework, and warranty costs?[1]

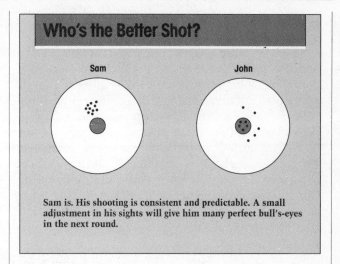

Who's the Better Shot?

Sam   John

Sam is. His shooting is consistent and predictable. A small adjustment in his sights will give him many perfect bull's-eyes in the next round.

That was precisely the reason. Imagine that in some Ford transmissions, many components near the *outer limits* of specified tolerances – that is, fine by the definitions of Zero Defects – were randomly assembled together. Then, many trivial deviations from the target tended to "stack up." An otherwise trivial variation in one part exacerbated a variation in another. Because of deviations, parts interacted with greater friction than they could withstand individually or with greater vibration than customers were prepared to endure.

Mazda managers worked consistently to bring parts in on target. Intuitively, they took a much more imaginative approach to on-line quality control than Ford managers did; they certainly grasped factory conformance in a way that superseded the pass/fail, in-spec/out-of-spec style of thinking associated with Zero Defects. Mazda managers worked on the assumption that robustness begins from meeting exact targets *consistently* – not from always staying within tolerances. They may not have realized this at the time, but they would have been even better off missing the target with perfect consistency than hitting it haphazardly – a point that is illuminated by this simple analogy:

Sam and John are at the range for target practice. After firing ten shots, they examine their targets. Sam has ten shots in a tight cluster just outside the bull's-eye circle. John, on the other hand, has five shots in the circle, but they are scattered all over it – as many near the perimeter as near dead center – and the rest of his shots are similarly dispersed around it (see the "Who's the Better Shot?" diagram).

Zero Defects theorists would say that John is the superior shooter because his performance betrays no

1. See Lance A. Ealey's admirable acccount of this case in *Quality by Design: Taguchi Methods® and U.S. Industry* (Dearborn, Mich.: ASI Press, 1988), pp. 61-62.

failures. But who would you really rather hire on as a bodyguard?

Sam's shooting is consistent and virtually predictable. He probably knows why he missed the circle completely. An adjustment to his sights will give many perfect bull's-eyes during the next round. John has a much more difficult problem. To reduce the dispersion of his shots, he must expose virtually all the factors under his control and find a way to change them in some felicitous combination. He may decide to change the position of his arms, the tightness of his sling, or the sequence of his firing: breathe, aim, slack, and squeeze. He will have little confidence that he will get all his shots in the bull's-eye circle next time around.

When extrapolated to the factory, a Sam-like performance promises greater product robustness. Once consistency is established—no mean feat, the product of relentless attention to the details of design and process both—adjusting performance to target is a simple matter: stack-up can be entirely obviated. If every drive shaft is .005 centimeters out, operators can adjust the position of the cutting tool. In the absence of consistent performance, getting more nearly on target can be terribly time-consuming.

But there is another side to this. There is a much higher probability of catastrophic stack-up from random deviations than from deviations that show consistency. Assuming that no part is grossly defective, a product made from parts that are all off target in exactly the same way is more likely to be robust than a product made from parts whose deviations are in-spec but unpredictable. We have statistical proofs of this, but a moment's reflection should be enough. If all parts are made consistently, the product will perform in a uniform way for customers and will be more easily perfected in the next version. If all parts are made erratically, some products will be perfect, and some will fall apart.

So the case against Zero Defects begins with this: Robustness derives from consistency. Where deviation is consistent, adjustment to the target is possible; catastrophic stack-up is more likely from scattered deviation within specifications than from consistent deviation outside. This regard for consistency, for being on target, has a fascinating and practical application.

The Quality Loss Function

Analysis of Ford's *overall* losses as compared with Mazda's suggests that when companies deviate from targets, they run an increasingly costly risk of loss.

Overall loss is quality loss plus factory loss. The more a manufacturer deviates from targets, the greater its losses.

From our experience, quality loss—the loss that comes after products are shipped—increases at a geometric rate. It can be roughly quantified as the Quality Loss Function (QLF), based on a simple quadratic formula. Loss increases by the square of deviation from the target value, $L = D^2C$, where the constant is determined by the cost of the countermeasure that the factory might use to get on target.

If you know what to do to get on target, then you know what this action costs per unit. If you balk at spending the money, then with every standard deviation from the target, you risk spending more and more. The greater the deviation from targets, the greater the compounded costs.

Let's say a car manufacturer chooses not to spend, say, $20 per transmission to get a gear exactly on target. QLF suggests that the manufacturer would wind up spending (when customers got mad) $80 for two standard deviations from the target ($20 multiplied by the square of two), $180 for three, $320 for four, and so forth.

This is a simple approximation, to be sure, not a law of nature. Actual field data cannot be expected to vindicate QLF precisely, and if your corporation has a more exacting way of tracking the costs of product failure, use it. But the tremendous value of QLF, apart from its bow to common sense, is that it translates the engineer's notion of deviation from targets into a simple cost estimate managers can use. QLF is especially helpful in the important early stages of new product development, when tolerances are set and quality targets are established.

Sony Televisions: Tokyo vs. San Diego

The compelling logic of QLF is best illustrated by the performance of Sony televisions in the late 1970s. The case demonstrates how engineering data and economic data can (and should) be seen in tandem.

Sony product engineers had ascertained that customers preferred pictures with a particular color density, let's call it a nominal density of 10. As color density deviated from 10, viewers became increasingly dissatisfied, so Sony set specification limits at no less than 7 and no more than 13.

Sony manufactured TV sets in two cities, San Diego and Tokyo. Sets shipped from San Diego were uniformly distributed within specs, which meant that a customer was as likely to buy a set with a color density of 12.6 as one with a density of 9.2. At the same time, a San Diego set was as likely to be near the corporate specification limits of 13 or 7 as near the customer satisfaction target of 10. Meanwhile, shipments from Tokyo tended to cluster near the target of 10, though at that time, about 3 out of every 1,000 sets actually fell outside of corporate standards.

> You gain nothing in shipping a product that barely satisfies corporate standards over one that just fails.

Akio Morita, the chairman of Sony, reflected on the discrepancy this way: "When we tell one of our Japanese employees that the measurement of a certain part must be within a tolerance of plus or minus five, for example, he will automatically strive to get that part as close to zero tolerance as possible. When we started our plant in the United States, we found that the workers would follow instructions perfectly. But if we said make it between plus or minus five, they would get it somewhere near plus or minus five all right, but rarely as close to zero as the Japanese workers did."

If Morita were to assign grades to the two factories' performances, he might say that Tokyo had many more As than San Diego, even if it did get a D now and then; 68% of Tokyo's production was in the A range, 28% in the B range, 4% in the C range, and 0.3% in the D range. Of course, San Diego made some out-of-spec sets; but it didn't ship its Fs. Tokyo shipped everything it built without bothering to check them. Should Morita have preferred Tokyo to San Diego?

The answer, remember, must be boiled down to dollars and cents, which is why the conventions of Zero Defects are of no use here. Suppose you bought a TV with a color density of 12.9, while your neighbor bought one with a density of 13.1. If you watch a program on his set, will you be able to detect any color difference between yours and his? Of course not. The color quality does not present a striking problem at the specification limit of 13. Things do not suddenly get more expensive for the San Diego plant if a set goes out at 13.1.

The losses start mounting when customers see sets at the target value of 10. Then, anything much away from 10 will seem unsatisfactory, and customers will demand visits from repairpeople or will demand replacement sets. Instead of spending a few dollars per set to adjust them close to targets, Sony would have to spend much more to make good on the sets – about two-thirds of the San Diego sets – that were actually displeasing customers. (Dissatisfaction certainly increases more between 11.5 and 13 than between 10 and 11.5.)

What Sony discovered is that you gain virtually nothing in shipping a product that just barely satisfies the corporate standard over a product that just fails. San Diego shipped marginal sets "without defects," but their marginal quality proved costly.

Using QLF, Sony might have come up with even more striking figures. Say the company estimated that the cost of the countermeasure required to put every set right – an assembly line countermeasure that puts every set at a virtual 10 – was $9. But for every San Diego set with a color density of 13 (three standard deviations from the target), Sony spent not $9 but $81. Total quality loss at San Diego should have been expected to be *three times* the total quality loss at the Tokyo factory.

Deviation: Signal to Noise

If Zero Defects doesn't work, what does? We have said that quality is mainly designed in, not controlled from without. In development work, engineers must discipline their decisions at virtually every step by comparing expected quality loss with known manufacturing cost. On the other hand, the reliability of QLF calculations is pretty obviously restricted by the accuracy of more preliminary measures. It is impossible to discern any loss function properly without first setting targets properly.

How *should* design engineers and manufacturing managers set targets? Let us proceed slowly, reconsidering what engineers do when they test components and subassemblies and how they establish what no particular part "wants to be" in the context of things that get in its way.

When Sony engineers designed their televisions, they assumed that discriminating customers would like a design that retained a good picture or "signal" far from the station, in a lightning storm, when the food processor was in use, and even when the power company was providing low voltage. Customers would be dismayed if the picture degraded every time they turned up the volume. They would reject a TV that developed snow and other annoying "noises" when afflicted by nasty operating conditions, which are themselves considered noises.

In our view, this metaphorical language – signal as compared with noise – can be used to speak of all

Orthogonal Arrays: Setting the Right Targets for Design

U.S. product engineers typically proceed by the "one factor at a time" method. A group of automotive-steering engineers—having identified 13 critical variables governing steering performance—would begin probing for design improvement by holding all variables at their current values and recording the result. In the second experiment, they would change just *one* of the variables—spring stiffness, say—to a lower or higher value, and if the result is an improvement (skidding at 40 mph, not 35), they will adopt *that* value as a design constant. Then comes the next experiment in which they'll change a different variable but not reconsider spring stiffness. They'll continue in this manner until they have nudged the steering system as close as possible to some ideal performance target (skidding at 50 mph, not 40).

The obvious trouble with such results is that they fail to take account of potentially critical interactions among variables—let alone the real variations in external conditions. While a certain spring stiffness provides ample performance when tire pressure is correct, how well will this stiffness work when tire pressure is too low or too high?

What these engineers need, therefore, is some efficient method to compare performance levels of all steering factors under test—and in a way that separates the *average* effect of spring stiffness at its high, low, and current settings on the various possible steering systems. The engineers could then select the spring-stiffness setting that consistently has the strongest positive effect on the best combination of variables.

If a particular spring-stiffness setting performs well in conjunction with each setting of all 12 other suspension factors, it stands a very good chance of reproducing positive results in the real world. This is an important advantage over the "one factor at a time" approach.

The orthogonal array can be thought of as a distillation mechanism through which the engineer's experimentation passes. Its great power lies in its ability to separate the effect each factor has on the average and the dispersion of the experiment as a whole. By exploiting this ability to sort out individual effects, engineers may track large numbers of factors simultaneously in each experimental run without confusion, thereby obviating the need to perform all possible combinations or to wait for the results of one experiment before proceeding with the next one.

Consider the orthogonal array for steering. Each of the rows in the array shown here constitutes one experiment, and each vertical column represents a single test factor. Column 1, for example, could represent spring stiffness. Engineers test each of the 13 steering factors at three different settings. (For spring stiffness, then, these would be the current setting, a stiffer setting, and a softer setting, notated 1, 2, and 3 in the array.)

Engineers perform 27 experiments on 13 variables, A through M. They run 27 experiments because they want to expose each performance level (1,2, and 3) to each other performance level an equal number of times (or 3 x 3 x 3). The engineer must perform all 27 experiments, adhering to the arrangement of the factor levels shown here and drawing lots in order to introduce an element of randomness to the experimentation.

products, not just televisions. The signal is what the product (or component or subassembly) is trying to deliver. Noises are the interferences that degrade signal, some of them coming from outside, some from complementary systems within the product. They are very much like the factors we spoke of as accounting for variations in product performance—environmental disturbances as well as disturbances engendered by the parts themselves.

And so it seems reasonable to define robustness as the virtue of a product with a high signal-to-noise ratio. Customers resent being told, "You were not expected to use our product in high humidity or in below-freezing temperatures." They want good performance under actual operating conditions—which are often less than perfect. We all assume that a product that performs better under adverse conditions will be that much more durable under normal conditions.

Signal-to-noise ratios are designed into products before the factory ramps up. The strength of a product's signal—hence, its robustness—is primarily the responsibility of the product designers. Good factories are faithful to the intention of the design. But mediocre designs will always result in mediocre products.

Choosing Targets: Orthogonal Arrays

How, then, do product designers maximize signal-to-noise ratios? World-class companies use a three-step decision-making process:

Factor Levels						
			Column			
No.	A	B C D	E F G	H I J	K L M	
1	1	1 1 1	1 1 1	1 1 1	1 1 1	Factor Levels
2	1	1 1 1	2 2 2	2 2 2	2 2 2	for First
3	1	1 1 1	3 3 3	3 3 3	3 3 3	Experiment
4	1	2 2 2	1 1 1	2 2 2	3 3 3	
5	1	2 2 2	2 2 2	3 3 3	1 1 1	
6	1	2 2 2	3 3 3	1 1 1	2 2 2	
7	1	3 3 3	1 1 1	3 3 3	2 2 2	
8	1	3 3 3	2 2 2	1 1 1	3 3 3	
9	1	3 3 3	3 3 3	2 2 2	1 1 1	
10	2	1 2 3	1 2 3	1 2 3	1 2 3	
11	2	1 2 3	2 3 1	2 3 1	2 3 1	
12	2	1 2 3	3 1 2	3 1 2	3 1 2	
13	2	2 3 1	1 2 3	2 3 1	3 1 2	
14	2	2 3 1	2 3 1	3 1 2	1 2 3	
15	2	2 3 1	3 1 2	1 2 3	2 3 1	
16	2	3 1 2	1 2 3	3 1 2	2 3 1	
17	2	3 1 2	2 3 1	1 2 3	3 1 2	
18	2	3 1 2	3 1 2	2 3 1	1 2 3	
19	3	1 3 2	1 3 2	1 3 2	1 3 2	
20	3	1 3 2	2 1 3	2 1 3	2 1 3	
21	3	1 3 2	3 2 1	3 2 1	3 2 1	
22	3	2 1 3	1 3 2	2 1 3	3 2 1	
23	3	2 1 3	2 1 3	3 2 1	1 3 2	
24	3	2 1 3	3 2 1	1 3 2	2 1 3	
25	3	3 2 1	1 3 2	3 2 1	2 1 3	Factor Levels
26	3	3 2 1	2 1 3	1 3 2	3 2 1	for 27th
27	3	3 2 1	3 2 1	2 1 3	1 3 2	Experiment

Experiments (label along vertical axis of rows)

Note that the three levels of factor A (1, 2, and 3) each are exposed to the three levels of the other 12 factors an equal number of times. This is true for all factors, A through M. Though not immediately obvious, the three levels of factor H, for example, are exposed equally, each appearing nine times for a total of 27.

It is important to note that each performance value is not exposed to every other possible combination of performance values. In this case, for instance, there is no experiment in which spring stiffness at 2 is exposed to all the other variables at 2; there is no row that shows 2 straight across. But if you look down each column, you will see that each performance level of each variable is exposed to each other variable *an equal number of times.* Before the experiment is completed, all three levels of spring stiffness will be exposed equally to all three levels of tire pressure, steering geometry, and so forth.

The engineer who uses an orthogonal array isn't primarily interested in the specific values yielded by the 27 experiments. By themselves, these values may not produce large improvements over existing performance. Instead, he or she is interested in distilling the effect that each of the three spring-stiffness settings has on the system as a whole. The "essence" of this approach is extracted along the array's vertical rather than horizontal axis.

The array allows the engineer to document results: those factor levels that will reduce performance variation as well as those that will guide the product back to its performance target once consistency has been achieved. The array also may be used to find variables that have little effect on robustness or target values: these can and should be set to their least costly levels.

–Lance A. Ealey

Lance A. Ealey, a member of the consulting staff of McKinsey & Co., is the author of Quality by Design: Taguchi Methods® and U.S. Industry *(ASI Press, 1988).*

1. They define the specific objective, selecting or developing the most appropriate signal and estimating the concomitant noise.

2. They define feasible options for the critical design values, such as dimensions and electrical characteristics.

3. They select the option that provides the greatest robustness or the greatest signal-to-noise ratio.

This is easier said than done, of course, which is why so many companies in Japan, and now in the United States, have moved to some form of simultaneous engineering. To define and select the correct signals and targets is no mean feat and requires the expertise of all product specialists. Product design, manufacturing, field support, and marketing—all of these should be worked out concurrently by an interfunctional team.

Product designers who have developed a "feel" for the engineering of particular products should take the lead in such teams. They can get away with only a few, limited experiments, where new people would have to perform many more. Progressive companies make an effort to keep their product specialists working on new versions rather than bump them up to management positions. Their compensation schemes reward people for doing what they do best.

But the virtues of teamwork beg the larger question of how to develop an efficient experimental strategy that won't drain corporate resources as you work to bring prototypes up to customer satisfaction. Intuition is not really an answer. Neither is interfunctionality or a theory of organization. Product designers need a scientific way to get at robustness. They have depended too long on art.

The most practical way to go about setting signal-to-noise ratios builds on the work of Sir Ronald Fisher, a British statistician whose brilliant contributions to agriculture are not much studied today. Most important is his strategy for systematic experimentation, including the astonishingly sensible plan known as the "orthogonal array."

Consider the complexity of improving a car's steering. Customers want it to respond consistently. Most engineers know that steering responsiveness depends on many critical design parameters—spring stiffness, shock absorber stiffness, dimensions of the steering and suspension mechanisms, and so on—all of which might be optimized to achieve the greatest possible signal-to-noise ratio.

It makes sense, moreover, to compare the initial design value to both a larger and a smaller value. If spring stiffness currently has a nominal value of 7, engineers may want to try the steering at 9 and at 5. One car engineer we've worked with established that there are actually 13 design variables for steering. If engineers were to compare standard, low, and high values for each critical variable, they would have 1,594,323 design options.

Proceed with intuition? Over a million possible permutations highlight the challenge—that of a blind search for a needle in a haystack—and steering is only one subsystem of the car. In Japan, managers say that engineers "like to fish with one rod"; engineers are optimistic that "the next cast will bring in the big fish"—one more experiment and they'll hit the ideal design. Naturally, repeated failure leads to more casts. The new product, still not robust to customers' conditions, is eventually forced into the marketplace by the pressures of time, money, and diminishing market share.

To complete the optimization of robustness most quickly, the search strategy must derive the maximum amount of information from a few trials. We won't go through the algebra here, but the key is to develop a system of trials that allows product engineers to analyze the *average* effect of change in factor levels under different sets of experimental conditions.

And this is precisely the virtue of the orthogonal array (see the insert, "Orthogonal Arrays: Setting the Right Targets for Design"). It balances the levels of performance demanded by customers against the many variables—or noises—affecting performance. An orthogonal array for 3 steering performance levels—low, medium, and high—can reduce the experimental possibilities to 27. Engineers might subject each of the 27 steering designs to some combination of noises, such as high/low tire pressure, rough/smooth road, high/low temperature. After all of the

trials are completed, signal-to-noise values may be used to select the best levels for each design variable.

If, for example, the average value for the first nine trials on spring stiffness is 32.4, then that could characterize level one of spring stiffness. If the average value for the second group of trials is 26.7, and the average for the third group 28.9, then we would select level one as the best value for spring stiffness. This averaging process is repeated to find the best level for each of the 13 design variables.

The orthogonal array is actually a sophisticated "switching system" into which many different design variables and levels of change can be plugged. This system was conceived to let the relatively inexperienced designer extract the *average* effect of each factor on the experimental results, so he or she can reach reliable conclusions despite the large number of changing variables.

Of course, once a product's characteristics are established so that a designer can say with certainty that design values—that is, optimized signal-to-noise ratios—do not interact at all, then orthogonal arrays are superfluous. The designer can instead proceed to test each design variable more or less independently, without concern for creating noise in other parts or subassemblies.

System Verification Test: The Moment of Truth

After they've maximized signal-to-noise ratios and optimized design values, engineers build prototypes. The robustness of the complete product is now verified in the System Verification Test (SVT)—perhaps the most critical event during product development.

In the SVT, the first prototypes are compared with the current benchmark product. Engineers subject both the prototype and the benchmark to the same extreme conditions they may encounter in actual use. Engineers also measure the same critical signal-to-noise ratios for all contenders. It is very important for the new product to surpass the robustness of the benchmark product. If the ideal nominal voltage is 115 volts, we want televisions that will have a signal of 10 even when voltage slips to a noisy 100 or surges to an equally noisy 130. Any deviations from the perfect signal must be considered in terms of QLF, that is, as a serious financial risk.

The robust product, therefore, is the one that minimizes the average of the square of the deviation from the target—averaged over the different customer-use conditions. Suppose you wish to buy a power supply and learn that you can buy one with a standard devia-

On-Line Quality: Shigeo Shingo's Shop Floor

Does tightening tolerances necessarily raise the specter of significantly higher production costs? Not according to Shigeo Shingo, the man who taught production engineering to a generation of Toyota managers. Now over 80, Shingo still actively promotes "Zero Quality Control," by which he aims to eliminate costly inspection processes or reliance on statistical quality control at the shop-floor level.

Shingo advocates the use of low-cost, in-process quality control mechanisms and routines that, in effect, incorporate 100% inspection at the source of quality problems. He argues for checking the causes rather than the effects of operator errors and machine abnormalities. This is achieved not through expensive automated control systems but through foolproofing methods such as *poka-yoke.*

Poka-yoke actually means "mistake proofing"; Shingo resists the idea that employees make errors because they are foolishly incompetent. Shingo believes all human beings have lapses in attention. The system, not the operator, is at fault when such defects occur. The poka-yoke method essentially builds the function of a checklist into an operation so we can never "forget what we have forgotten."

For example, if an operator needs to insert a total of nine screws into a subassembly, modify the container so it releases nine at a time. If a screw remains, the operator knows the operation is not complete. Many poka-yoke ideas are based on familiar foolproofing concepts; the mechanisms are also related in principle to *jidohka,* or autonomation—the concept of low-cost "intelligent machines" that stop automatically when processing is completed or when an abnormality occurs.

Shingo recommends four principles for implementing poka-yoke:

1. *Control upstream, as close to the source of the potential defect as possible.* For example, modify the form of a symmetrical workpiece just slightly to assure correct positioning with a jig or sensor keyed to an asymmetrical characteristic. Also, attach a monitoring device that will detect a material abnormality or an abnormal machine condition and will trigger shutdown before a defect is generated and passed on to the next process.

2. *Establish controls in relation to the severity of the problem.* A simple signal or alarm may be sufficient to check an error that is easily corrected by the operator, but preventing further progress until the error is corrected is even better. For example, a control-board counter counts the number of spot welds performed and operates jig clamps; if one is omitted, the workpiece cannot be removed from the jig until the error is corrected.

3. *Think smart and small.* Strive for the simplest, most efficient, and most economical intervention. Don't overcontrol—if operator errors result from a lack of operations, improve methods before attempting to control the results. Similarly, if the cost of equipment adjustment is high, improve equipment reliability and consider how to simplify adjustment operations before implementing a costly automated-inspection system.

4. *Don't delay improvement by overanalyzing.* Poka-yoke solutions are usually the product of decisiveness and quick action on the shop floor. While design improvements can reduce manufacturing defects in the long run, you can implement many poka-yoke ideas at very low cost within hours of their conception, effectively closing the quality gap until you develop a more robust design.

Developed cooperatively by operators, production engineers, and machine-shop personnel, poka-yoke methods are employed extensively in processing and assembly operations in Japan and represent one of the creative pinnacles of continuous shop-floor improvement. In its most developed state, such improvement activity can support off-line quality engineering efforts by feeding back a continuous flow of data about real, in-process quality problems.

—Connie Dyer

Connie Dyer is senior editor at Productivity Press. She worked with Shigeo Shingo on, among other books, Zero Quality Control: Source Inspection and the Poka-Yoke System *(Productivity Press, 1986).*

tion of one volt. Should you take it? If the mean value of output voltage is 1,000 volts, most people would think that, on average, an error of only one volt is very good. However, if the average output were 24 volts, then a standard deviation of one seems very large. We must always consider the ratio of the mean value divided by the standard deviation.

The SVT gives a very strong indication, long before production begins, of whether customers will perceive the new product as having world-class quality and performance. After the new design is verified to have superior robustness, engineers may proceed to solve routine problems, fully confident that the product will steadily increase customer loyalty.

Back to the Factory

The relationship of field to factory proves to be a subtle one – the converse of what one might expect. We know that if you control for variance in the factory, you reduce failure in the field. But as we said at the outset, a concerted effort to reduce product failure in the field will simultaneously reduce the number of defectives in the factory.

Strive to reduce variances in the components of the product and you will reduce variances in the production system as a whole. Any strengthening of a design – that is, any marked increase of a product's signal-to-noise ratio – will simultaneously reduce a factory's quality losses.

Why should this be so? For many reasons, most importantly the symmetries between design for robustness and design for manufacture. Think of how much more robust products have become since the introduction of molded plastics and solid-state circuitry. Instead of serving up many interconnected wires and tubes and switches – any one of which can fail – engineers can now imprint a million transistors on a virtually indestructible chip. Instead of joining many components together with screws and fasteners, we can now consolidate parts into subassemblies and mount them on molded frames that snap together.

All of these improvements greatly reduce opportunities for noise interfering with signal; they were developed to make products robust. Yet they also have made products infinitely more manufacturable. The principles of designing for robustness are often indistinguishable from the principles of designing for manufacture – reduce the number of parts, consolidate subsystems, integrate the electronics.

A robust product can tolerate greater variations in the production system. Please the customer and you will please the manufacturing manager. Prepare for variances in the field and you will pave the way for reducing variations on the shop floor. None of this means the manufacturing manager should stop trying to reduce process variations or to achieve the same variations with faster, cheaper processes. And there are obvious exceptions proving the rule – chip production, for example, where factory controls are ever more stringent – though it is hard to think of exceptions in products such as cars and consumer electronics.

The factory is a place where workers must try to meet, not deviate from, the nominal targets set for products. It is time to think of the factory as a product with targets of its own. Like a product, the factory may be said to give off an implicit signal – the consistent production of robust products – and to be subject to the disruptions of noise – variable temperatures, degraded machines, dust, and so forth. Using QLF, choices in the factory, like choices for the product, can be reduced to the cost of deviation from targets.

Consider, for instance, that a cylindrical grinder creates a cylindrical shape more consistently than a lathe. Product designers have argued for such dedicated machines; they want the greatest possible precision. Manufacturing engineers have traditionally favored the less precise lathe because it is more flexible and it reduces production cost. Should management favor the more precise cylindrical grinder? How do you compare each group's choice with respect to quality loss?

In the absence of QLF's calculation, the most common method for establishing manufacturing tolerances is to have a concurrence meeting. Design engineers sit on one side of the conference room, manufacturing engineers on the opposite side. The product design engineers start chanting "Tighter Tolerance, Tighter Tolerance, Tighter Tolerance," and the manufacturing engineers respond with "Looser Tolerance, Looser Tolerance, Looser Tolerance." Presumably, the factory would opt for a lathe if manufacturing chanted louder and longer. But why follow such an irrational process when product design people and manufacturing people *can* put a dollar value on quality precision?

Management should choose the precision level that minimizes the total cost, production cost plus quality loss – the basics of QLF. Managers can compare the costs of competing factory processes by adding the manufacturing cost and the average quality loss (from expected deviations) of each process. They gain economical precision by evaluating feasible alternative production processes, such as the lathe and cylindrical grinder. What would be the quality loss if the factory used the lathe? Are the savings worth the future losses?

Similar principles may be applied to larger systems. In what may be called "process parameter design," manufacturers can optimize production parameters – spindle speed, depth of cut, feed rate, pressure, temperature – according to an orthogonal array, much like the spring stiffness in a steering mechanism. Each row of the orthogonal array may define a different production trial. In each trial, engineers produce and measure several parts and then use the data to calculate the signal-to-noise ratio for that trial. In a final step, they establish the best value for each production parameter.

The result? A robust process – one that produces improved uniformity of parts and often enables managers to simultaneously speed up production and reduce cycle time.

How Much Intervention?

Finally, there is the question of how much to intervene *during* production.

Take the most common kind of intervention, on-line checking and adjusting of machinery and process. In the absence of any operator monitoring, parts tend to deviate progressively from the target. Without guidance, different operators have widely varying notions of (1) how often they should check their machines and (2) how big the discrepancy must be before they adjust the process to bring the part value back near the target.

By applying QLF, you can standardize intervention. The cost of checking and adjusting has always been easy to determine; you simply have to figure the cost of downtime. With QLF, managers can also figure the cost of *not* intervening, that is, the dollar value to the company of reduced parts variation.

Let's go back to drive shafts. The checking interval is three, and the best adjustment target is 1/1,000th of an inch. If the measured discrepancy from the target is less than 1/1,000th of an inch, production continues. If the measured discrepancy exceeds this, the process is adjusted back to the target. Does this really enable operators to keep the products near the target in a way that minimizes total cost?

It might be argued that measuring every third shaft is too expensive. Why not every tenth? There is a way to figure this out. Say the cost of intervention is 30 cents, and shafts almost certainly deviate from the target value every fifth or sixth operation. Then, out of every ten produced, at least four bad shafts will go out, and quality losses will mount. If the seventh shaft comes out at two standard deviations, the cost will be $1.20; if the tenth comes out at three standard

deviations, the cost will be $2.70; and so on. Perhaps the best interval to check is every fourth shaft or every fifth, not every third. If the fourth shaft is only one standard deviation from the target value, intervention is probably not worth the cost.

The point, again, is that these things can and should be calculated. There isn't any reason to be fanatical about quality if you *cannot* justify your fanaticism by QLF. Near the target, production should continue without adjustment; the quality loss is small. Outside the limit, the process should be adjusted before production continues.

This basic approach to intervention can also be applied to preventive maintenance. Excessive preventive maintenance costs too much. Inadequate preventive maintenance will increase quality loss excessively. Optimized preventive maintenance will minimize total cost.

In Japan, it is said that a manager who trades away quality to save a little manufacturing expense is "worse than a thief" – a little harsh, perhaps, but plausible. When a thief steals $200 from your company, there is no net loss in wealth between the two of you, just an exchange of assets. Decisions that create huge quality losses throw away social productivity, the wealth of society.

QLF's disciplined, quantitative approach to quality builds on and enhances employee involvement activities to improve quality and productivity. Certainly, factory-focused improvement activities do not by and large increase the robustness of a product. They can help realize it, however, by reducing the noise generated by the complex interaction of shop-floor quality factors – operators, operating methods, equipment, and material.

Employees committed to hitting the bull's-eye consistently cast a sharper eye on every feature of the factory environment. When their ingenuity and cost-consciousness are engaged, conditions change dramatically, teams prosper, and valuable data proliferate to support better product and process design. An early, companywide emphasis on robust product design can even reduce development time and smooth the transition to full-scale production.

Too often managers think that quality is the responsibility of only a few quality control people off in a factory corner. It should be evident by now that quality is for everyone, most of all the business's strategists. It is only through the efforts of every employee, from the CEO on down, that quality will become second nature. The most elusive edge in the new global competition is the galvanizing pride of excellence.

Reprint 90114

Design is a team effort, but how do marketing and engineering talk to each other?

The House of Quality

by JOHN R. HAUSER and DON CLAUSING

Digital Equipment, Hewlett-Packard, AT&T, and ITT are getting started with it. Ford and General Motors use it—at Ford alone there are more than 50 applications. The "house of quality," the basic design tool of the management approach known as quality function deployment (QFD), originated in 1972 at Mitsubishi's Kobe shipyard site. Toyota and its suppliers then developed it in numerous ways. The house of quality has been used successfully by Japanese manufacturers of consumer electronics, home appliances, clothing, integrated circuits, synthetic rubber, construction equipment, and agricultural engines. Japanese designers use it for services like swimming schools and retail outlets and even for planning apartment layouts.

A set of planning and communication routines, quality function deployment focuses and coordinates skills within an organization, first to design, then to manufacture and market goods that custom-ers want to purchase and will continue to purchase. The foundation of the house of quality is the belief that products should be designed to reflect custom-ers' desires and tastes—so marketing people, design engineers, and manufacturing staff must work closely together from the time a product is first conceived.

The house of quality is a kind of conceptual map that provides the means for interfunctional planning and communications. People with different prob-

John R. Hauser, at the Harvard Business School as a Marvin Bower fellow during the current academic year, is professor of management science at MIT's Sloan School of Management. He is the author, with Glen L. Urban, of Design & Marketing of New Products *(Prentice-Hall, 1980). Don Clausing is Bernard M. Gordon Adjunct Professor of Engineering Innovation and Practice at MIT. Previously he worked for Xerox Corporation. He introduced QFD to Ford and its supplier companies in 1984.*

lems and responsibilities can thrash out design priorities while referring to patterns of evidence on the house's grid.

What's So Hard About Design

David Garvin points out that there are many dimensions to what a consumer means by quality and that it is a major challenge to design products that satisfy all of these at once.[1] Strategic quality management means more than avoiding repairs for consumers. It means that companies learn from customer experience and reconcile what they want with what engineers can reasonably build.

Before the industrial revolution, producers were close to their customers. Marketing, engineering, and manufacturing were integrated – in the same individual. If a knight wanted armor, he talked directly to the armorer, who translated the knight's desires into a product. The two might discuss the material – plate rather than chain armor – and details like fluted surfaces for greater bending strength. Then the armorer would design the production process. For strength – who knows why? – he cooled the steel plates in the urine of a black goat. As for a production plan, he arose with the cock's crow to light the forge fire so that it would be hot enough by midday.

Today's fiefdoms are mainly inside corporations. Marketing people have their domain, engineers theirs. Customer surveys will find their way onto designers' desks, and R&D plans reach manufacturing engineers. But usually, managerial functions remain disconnected, producing a costly and demoralizing environment in which product quality and the quality of the production process itself suffer.

Top executives are learning that the use of interfunctional teams benefits design. But if top management *could* get marketing, designing, and manufacturing executives to sit down together, what should these people talk about? How could they get their meeting off the ground? This is where the house of quality comes in.

Consider the location of an emergency brake lever in one American sporty car. Placing it on the left between the seat and the door solved an engineering problem. But it also guaranteed that women in skirts could not get in and out gracefully. Even if the system were to last a lifetime, would it satisfy customers?

In contrast, Toyota improved its rust prevention record from one of the worst in the world to one of the best by coordinating design and production decisions to focus on this customer concern. Using the house of quality, designers broke down "body durability" into 53 items covering everything from climate to modes of operation. They obtained customer evaluations and ran experiments on nearly every detail of production, from pump operation to temperature control and coating composition. Decisions on sheet metal details, coating materials, and baking temperatures were all focused on those aspects of rust prevention most important to customers.

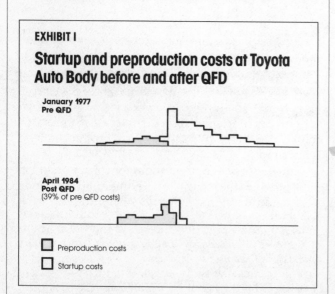

EXHIBIT I

Startup and preproduction costs at Toyota Auto Body before and after QFD

January 1977
Pre QFD

April 1984
Post QFD
(39% of pre QFD costs)

☐ Preproduction costs
☐ Startup costs

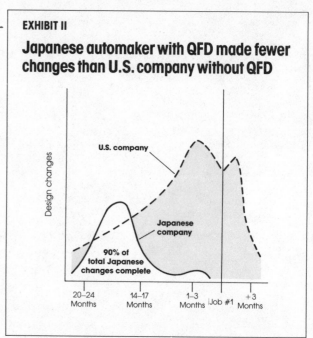

EXHIBIT II

Japanese automaker with QFD made fewer changes than U.S. company without QFD

Design changes

U.S. company

Japanese company

90% of total Japanese changes complete

20–24 Months   14–17 Months   1–3 Months Job #1   +3 Months

Source for Exhibits I and II: Lawrence P. Sullivan, "Quality Function Deployment," *Quality Progress,* June 1986, p. 39. © 1986 American Society for Quality Control. Reprinted by permission.

1. David A. Garvin, "Competing on the Eight Dimensions of Quality," HBR November-December 1987, p. 101.

Today, with marketing techniques so much more sophisticated than ever before, companies can measure, track, and compare customers' perceptions of products with remarkable accuracy; all companies have opportunities to compete on quality. And costs certainly justify an emphasis on quality design. By looking first at customer needs, then designing across corporate functions, manufacturers can reduce prelaunch time and after-launch tinkering.

Exhibit I compares startup and preproduction costs at Toyota Auto Body in 1977, before QFD, to those costs in 1984, when QFD was well under way. House of quality meetings early on reduced costs by more than 60%. *Exhibit II* reinforces this evidence by comparing the number of design changes at a Japanese auto manufacturer using QFD with changes at a U.S. automaker. The Japanese design was essentially frozen before the first car came off the assembly line, while the U.S. company was still revamping months later.

Building the House

There is nothing mysterious about the house of quality. There is nothing particularly difficult about it either, but it does require some effort to get used to its conventions. Eventually one's eye can bounce knowingly around the house as it would over a roadmap or a navigation chart. We have seen some applications that started with more than 100 customer requirements and more than 130 engineering considerations. A fraction of one subchart, in this case for the door of an automobile, illustrates the house's basic concept well. We've reproduced this subchart portion in the illustration "House of Quality," and we'll discuss each section step-by-step.

 What do customers want? The house of quality begins with the customer, whose requirements are called customer attributes (CAs)—phrases customers use to describe products and product characteristics (see *Exhibit III*). We've listed a few here; a typical application would have 30 to 100 CAs. A car door is "easy to close" or "stays open on a hill"; "doesn't leak in rain" or allows "no (or little) road noise." Some

EXHIBIT III

Customer attributes and bundles of CAs for a car door

PRIMARY	SECONDARY	TERTIARY
Good operation and use	EASY TO OPEN AND CLOSE DOOR	Easy to close from outside
		Stays open on a hill
		Easy to open from outside
		Doesn't kick back
		Easy to close from inside
		Easy to open from inside
	ISOLATION	Doesn't leak in rain
		No road noise
		Doesn't leak in car wash
		No wind noise
		Doesn't drip water or snow when open
		Doesn't rattle
	ARM REST	Soft, comfortable
		In right position
Good appearance	INTERIOR TRIM	Material won't fade
		Attractive (nonplastic look)
	CLEAN	Easy to clean
		No grease from door
	FIT	Uniform gaps between matching panels

Japanese companies simply place their products in public areas and encourage potential customers to examine them, while design team members listen and note what people say. Usually, however, more formal market research is called for, via focus groups, in-depth qualitative interviews, and other techniques.

CAs are often grouped into bundles of attributes that represent an overall customer concern, like "open-close" or "isolation." The Toyota rust-prevention study used eight levels of bundles to get from the total car down to the car body. Usually the project team groups CAs by consensus, but some companies are experimenting with state-of-the-art research techniques that derive groupings directly from customers' responses (and thus avoid arguments in team meetings).

CAs are generally reproduced in the customers' own words. Experienced users of the house of quality try to preserve customers' phrases and even clichés—knowing that they will be translated simultaneously by product planners, design engineers, manufacturing engineers, and salespeople. Of course, this raises the problem of interpretation: What does a customer really mean by "quiet" or "easy"? Still, designers' words and inferences may correspond even less to customers' actual views and can therefore mislead teams into tackling problems customers consider unimportant.

Not all customers are end users, by the way. CAs can include the demands of regulators ("safe in a side collision"), the needs of retailers ("easy to display"),

---

EXHIBIT IV

Relative-importance weights of customer attributes

BUNDLES	CUSTOMER ATTRIBUTES	RELATIVE IMPORTANCE
EASY TO OPEN AND CLOSE DOOR	Easy to close from outside	7
	Stays open on a hill	5
ISOLATION	Doesn't leak in rain	3
	No road noise	2

A complete list totals 100%

---

the requirements of vendors ("satisfy assembly and service organizations"), and so forth.

Are all preferences equally important? Imagine a good door, one that is easy to close and has power windows that operate quickly. There is a problem, however. Rapid operation calls for a bigger motor, which makes the door heavier and, possibly, harder to close. Sometimes a creative solution can be found that satisfies all needs. Usually, however, designers have to trade off one benefit against another.

To bring the customer's voice to such deliberations, house of quality measures the relative importance to the customer of all CAs. Weightings are based on team members' direct experience with customers or on surveys. Some innovative businesses are using statistical techniques that allow customers to state their preferences with respect to existing and hypothetical products. Other companies use "revealed preference techniques," which judge consumer tastes by their actions as well as by their words–an approach that is more expensive and difficult to perform but yields more accurate answers. (Consumers say that avoiding sugar in cereals is important, but do their actions reflect their claims?)

Weightings are displayed in the house next to each CA–usually in terms of percentages, a complete list totaling 100% (see *Exhibit IV*).

Will delivering perceived needs yield a competitive advantage? Companies that want to match or exceed their competition must first know where they stand relative to it. So on the right side of the house, opposite the CAs, we list customer evaluations of competitive cars matched to "our own" (see *Exhibit V*).

Ideally, these evaluations are based on scientific surveys of customers. If various customer segments

evaluate products differently–luxury vs. economy car buyers, for example–product-planning team members get assessments for each segment.

Comparison with the competition, of course, can identify opportunities for improvement. Take our car door, for example. With respect to "stays open on a hill," every car is weak, so we could gain an advantage here. But if we looked at "no road noise" for the same automobiles, we would see that we already have an advantage, which is important to maintain.

Marketing professionals will recognize the right-hand side of *Exhibit V* as a "perceptual map." Perceptual maps based on bundles of CAs are often used to identify strategic positioning of a product or product line. This section of the house of quality provides a natural link from product concept to a company's strategic vision.

How can we change the product? The marketing domain tells us what to do, the engineering domain tells us how to do it. Now we need to describe the product in the language of the engineer. Along the top of the house of quality, the design team lists those engineering characteristics (ECs) that are likely to affect one or more of the customer attributes (see *Exhibit VI*). The negative sign on "energy to close door" means engineers hope to reduce the energy required. If a standard engineering characteristic affects no CA, it may be redundant to the EC list on the house, or the team may have missed a customer attribute. A CA unaffected by any EC, on the other hand, presents opportunities to expand a car's physical properties.

Any EC may affect more than one CA. The resistance of the door seal affects three of the four customer attributes shown in *Exhibit VI*–and others shown later.

Engineering characteristics should describe the product in measurable terms and should directly affect customer perceptions. The weight of the door will be *felt* by the customer and is therefore a relevant EC. By contrast, the thickness of the sheet metal is a part characteristic that the customer is unlikely to perceive directly. It affects customers only by influencing the weight of the door and other engineering characteristics, like "resistance to deformation in a crash."

In many Japanese projects, the interfunctional team begins with the CAs and generates measurable characteristics for each, like foot-pounds of energy required to close the door. Teams should avoid ambiguity in interpretation of ECs or hasty justification of current quality control measurement practices. This is a time for systematic, patient analysis of each characteristic, for brainstorming. Vagueness will eventually yield indifference to things customers need. Characteristics that are trivial will make the team lose sight of the overall design and stifle creativity.

How much do engineers influence customer-perceived qualities? The interfunctional team now fills in the body of the house, the "relationship matrix," indicating how much each engineering characteristic affects each customer attribute. The team seeks consensus on these evaluations, basing them on expert engineering experience, customer responses, and tabulated data from statistical studies or controlled experiments.

The team uses numbers or symbols to establish the strength of these relationships (see *Exhibit VII*). Any symbols will do; the idea is to choose those that work best. Some teams use red symbols for relationships based on experiments and statistics and pencil marks for relationships based on judgment or intuition. Others use numbers from statistical studies. In our house, we use check marks for positive and crosses for negative relationships.

Once the team has identified the voice of the customer and linked it to engineering characteristics, it adds objective measures at the bottom of the house beneath the ECs to which they pertain (see *Exhibit VIII*). When objective measures are known, the team can eventually move to establish target values—ideal new measures for each EC in a redesigned product. If the team did its homework when it first identified the ECs, tests to measure benchmark values should be easy to complete. Engineers determine the relevant units of measurement—foot-pounds, decibels, etc.

Incidentally, if customer evaluations of CAs do not correspond to objective measures of related ECs —if, for example, the door requiring the least energy to open is perceived as "hardest to open"—then perhaps the measures are faulty or the car is suffering from an image problem that is skewing consumer perceptions.

How does one engineering change affect other characteristics? An engineer's change of the gear ratio on a car window may make the window motor smaller but the window go up more slowly. And if the engineer enlarges or strengthens the mechanism, the door probably will be heavier, harder to open, or may be less prone to remain open on a slope. Of course,

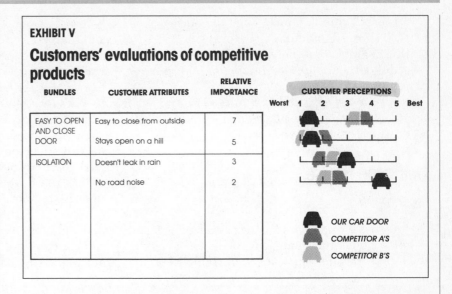

EXHIBIT V

Customers' evaluations of competitive products

BUNDLES	CUSTOMER ATTRIBUTES	RELATIVE IMPORTANCE	CUSTOMER PERCEPTIONS
EASY TO OPEN AND CLOSE DOOR	Easy to close from outside	7	
	Stays open on a hill	5	
ISOLATION	Doesn't leak in rain	3	
	No road noise	2	

OUR CAR DOOR
COMPETITOR A'S
COMPETITOR B'S

there might be an entirely new mechanism that improves all relevant CAs. Engineering is creative solutions and a balancing of objectives.

The house of quality's distinctive roof matrix helps engineers specify the various engineering features that have to be improved collaterally (see *Exhibit IX*). To improve the window motor, you may have to improve the hinges, weather stripping, and a range of other ECs.

Sometimes one targeted feature impairs so many others that the team decides to leave it alone. The roof matrix also facilitates necessary engineering trade-offs. The foot-pounds of energy needed to close the door, for example, are shown in negative relation to "door seal resistance" and "road noise reduction." In many ways, the roof contains the most critical information for engineers because they use it to balance the trade-offs when addressing customer benefits.

Incidentally, we have been talking so far about the basics, but design teams often want to ruminate on other information. In other words, they custom-build their houses. To the column of CAs, teams may add other columns for histories of customer complaints. To the ECs, a team may add the costs of servicing these complaints. Some applications add data from the sales force to the CA list to represent strategic marketing decisions. Or engineers may add a row that indicates the degree of technical difficulty, showing in their own terms how hard or easy it is to make a change.

Some users of the house impute relative weights to the engineering characteristics. They'll establish that the energy needed to close the door is roughly twice as important to consider as, say, "check force on 10° slope." By comparing weighted characteristics to actual component costs, creative design teams set priorities for improving components. Such information is particularly important when cost cutting is a goal. (*Exhibit X* includes rows for technical difficulty, imputed importance of ECs, and estimated costs.)

There are no hard-and-fast rules. The symbols, lines, and configurations that work for the particular team are the ones it should use.

Using the House

How does the house lead to the bottom line? There is no cookbook procedure, but the house helps the team to set targets, which are, in fact, entered on the bottom line of the house. For engineers it is a way to summarize basic data in usable form. For marketing executives it represents the customer's voice. General managers use it to discover strategic opportunities. Indeed, the house encourages all of these groups to work together to understand one another's priorities and goals.

The house relieves no one of the responsibility of making tough decisions. It does provide the means for all participants to debate priorities.

Let's run through a couple of hypothetical situations to see how a design team uses the house.

■ Look at *Exhibit X*. Notice that our doors are much more difficult to close from the outside than those on competitors' cars. We decide to look further because our marketing data say this customer attribute is

EXHIBIT VI

Engineering characteristics tell how to change the product

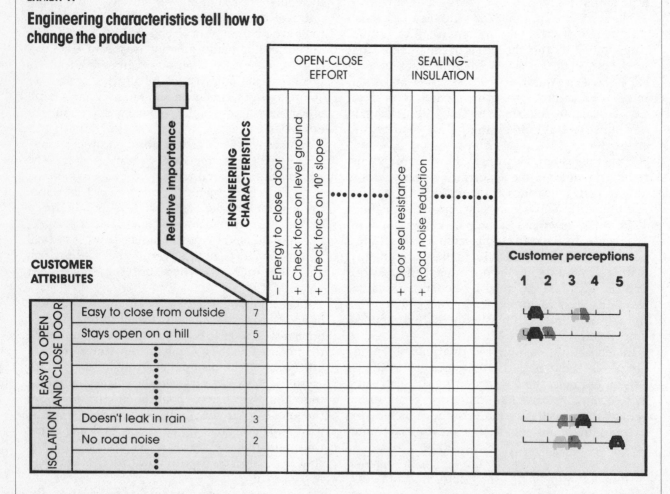

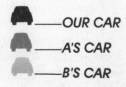

EXHIBIT VII

Relationship matrix shows how engineering decisions affect customer perceptions

CUSTOMER ATTRIBUTES		Relative importance	Energy to close door (−)	Check force on level ground (+)	Check force on 10° slope (+)				Door seal resistance (+)	Road noise reduction (+)			
EASY TO OPEN AND CLOSE DOOR	Easy to close from outside	7	✔						✗				
	Stays open on a hill	5		✔	✔								
	⋮												
ISOLATION	Doesn't leak in rain	3							✔				
	No road noise	2							✔	✔			
	⋮												

Relationships

✔ Strong positive
✔ Medium positive
✗ Medium negative
✗ Strong negative

Customer perceptions

1 2 3 4 5

——OUR CAR
——A'S CAR
——B'S CAR

important. From the central matrix, the body of the house, we identify the ECs that affect this customer attribute: energy to close door, peak closing force, and door seal resistance. Our engineers judge the energy to close the door and the peak closing force as good candidates for improvement together because they are strongly, positively related to the consumer's desire to close the door easily. They determine to consider all the engineering ramifications of door closing.

Next, in the roof of the house, we identify which other ECs might be affected by changing the door closing energy. Door opening energy and peak closing force are positively related, but other ECs (check force on level ground, door seals, window acoustic transmission, road noise reduction) are bound to be changed in the process and are negatively related. It is not an easy decision. But with objective measures of competitors' doors, customer perceptions, and considering information on cost and technical difficulty, we—marketing people, engineers, and top managers—decide that the benefits outweigh the costs. A new door closing target is set for our door—7.5 foot-pounds of energy. This target, noted on the very bottom of the house directly below the relevant EC, establishes the goal to have the door "easiest to close."

■ Look now at the customer attribute "no road noise" and its relationship to the acoustic transmission of the window. The "road noise" CA is only mildly important to customers, and its relationship to the specifications of the window is not strong. Window design will help only so much to keep things quiet. Decreasing the acoustic transmission usually makes the window heavier. Examining the roof of the house, we see that more weight would have a negative impact on ECs (open-close energy, check forces, etc.) that, in turn, are strongly related to CAs that are more important to the customer than quiet ("easy to close," "stays open on a hill"). Finally, marketing data show that we already do well on road noise; customers perceive our car as better than competitors'.

In this case, the team decides not to tamper with the window's transmission of sound. Our target stays equal to our current acoustic values.

In setting targets, it is worth noting that the team should emphasize customer-satisfaction values and not emphasize tolerances. Do not specify "between 6 and 8 foot-pounds," but rather say, "7.5 foot-pounds." This may seem a small matter, but it is important. The rhetoric of tolerances encourages drift toward the least costly end of the specification limit and

EXHIBIT VIII

Objective measures evaluate competitive products

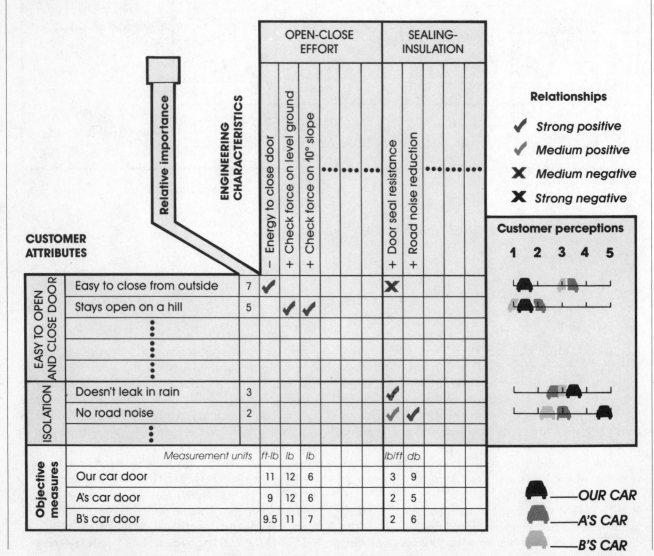

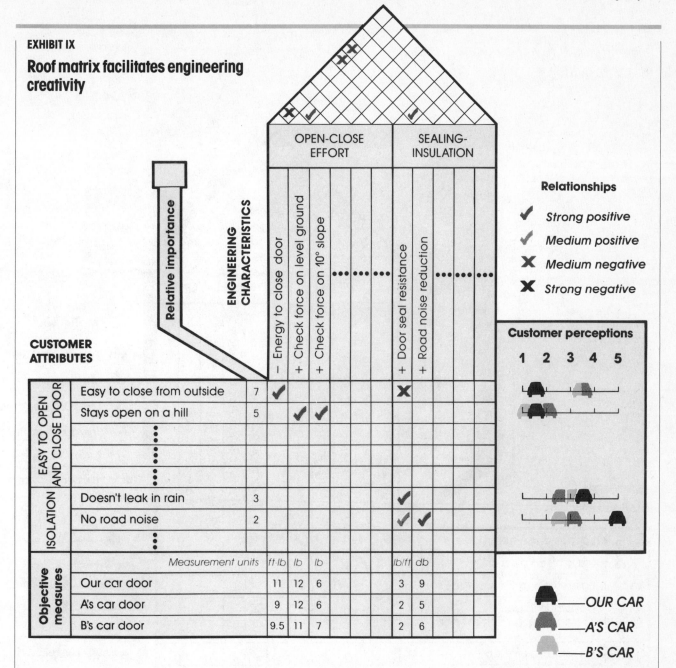

EXHIBIT IX

Roof matrix facilitates engineering creativity

CUSTOMER ATTRIBUTES		Relative importance	OPEN-CLOSE EFFORT						SEALING-INSULATION				
			− Energy to close door	+ Check force on level ground	+ Check force on 10° slope				+ Door seal resistance	+ Road noise reduction			
EASY TO OPEN AND CLOSE DOOR	Easy to close from outside	7	✔						✖				
	Stays open on a hill	5		✔	✔								
ISOLATION	Doesn't leak in rain	3							✔				
	No road noise	2							✔	✔			
Objective measures	Measurement units		ft·lb	lb	lb				lb/ft	db			
	Our car door		11	12	6				3	9			
	A's car door		9	12	6				2	5			
	B's car door		9.5	11	7				2	6			

ENGINEERING CHARACTERISTICS

Relationships

✔ Strong positive
✓ Medium positive
✖ Medium negative
✖ Strong negative

Customer perceptions

1 2 3 4 5

——OUR CAR
——A'S CAR
——B'S CAR

does not reward designs and components whose engineering values closely attain a specific customer-satisfaction target.

The Houses Beyond

The principles underlying the house of quality apply to any effort to establish clear relations between manufacturing functions and customer satisfaction that are not easy to visualize. Suppose that our team

decides that doors closing easily is a critical attribute and that a relevant engineering characteristic is closing energy. Setting a target value for closing energy gives us a goal, but it does not give us a door. To get a door, we need the right parts (frame, sheet metal, weather stripping, hinges, etc.), the right processes to manufacture the parts and assemble the product, and the right production plan to get it built.

If our team is truly interfunctional, we can eventually take the "hows" from our house of quality and make them the "whats" of another house, one mainly concerned with detailed product design. En-

EXHIBIT X

House of quality

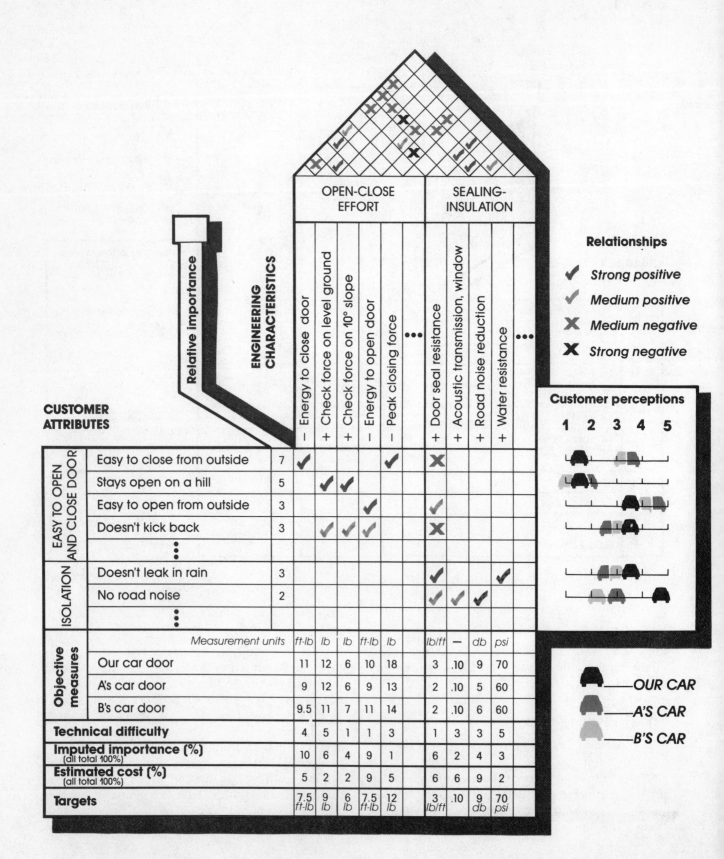

Relationships

✔ Strong positive
✔ Medium positive
✘ Medium negative
✘ Strong negative

Customer perceptions

	ENGINEERING CHARACTERISTICS		OPEN-CLOSE EFFORT					SEALING-INSULATION			
	Relative importance		− Energy to close door	+ Check force on level ground	+ Check force on 10° slope	− Energy to open door	− Peak closing force	+ Door seal resistance	− Acoustic transmission, window	+ Road noise reduction	+ Water resistance

CUSTOMER ATTRIBUTES

EASY TO OPEN AND CLOSE DOOR	Easy to close from outside	7	✔				✔	✘			
	Stays open on a hill	5		✔	✔						
	Easy to open from outside	3			✔			✔			
	Doesn't kick back	3		✔	✔	✔		✘			
ISOLATION	Doesn't leak in rain	3						✔			✔
	No road noise	2						✔	✔	✔	

Objective measures	Measurement units	ft-lb	lb	lb	ft-lb	lb	lb/ft	−	db	psi
	Our car door	11	12	6	10	18	3	.10	9	70
	A's car door	9	12	6	9	13	2	.10	5	60
	B's car door	9.5	11	7	11	14	2	.10	6	60

Technical difficulty	4	5	1	1	3	1	3	3	5
Imputed importance (%) (all total 100%)	10	6	4	9	1	6	2	4	3
Estimated cost (%) (all total 100%)	5	2	2	9	5	6	6	9	2
Targets	7.5 ft-lb	9 lb	6 lb	7.5 ft-lb	12 lb	3 lb/ft	.10	9 db	70 psi

OUR CAR
A'S CAR
B'S CAR

EXHIBIT XI

Linked houses convey the customer's voice through to manufacturing

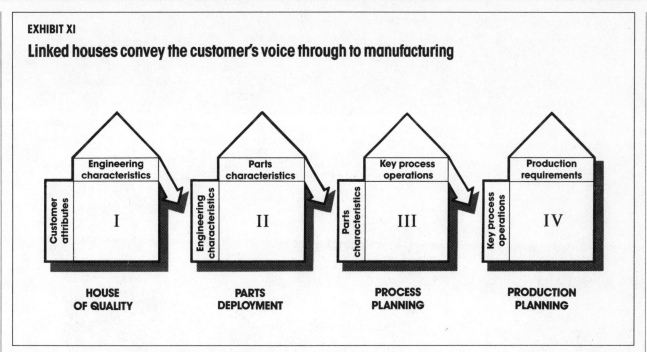

Source: Modified from a figure supplied by the American Supplier Institute, Inc., Dearborn, Michigan.

gineering characteristics like foot-pounds of closing energy can become the rows in a parts deployment house, while parts characteristics – like hinge properties or the thickness of the weather stripping – become the columns (see *Exhibit XI*).

This process continues to a third and fourth phase as the "hows" of one stage become the "whats" of the next. Weather-stripping thickness – a "how" in the parts house – becomes a "what" in a process planning house. Important process operations, like "rpm of the extruder producing the weather stripping" become the "hows." In the last phase, production planning, the key process operations, like "rpm of the extruder," become the "whats," and production requirements – knob controls, operator training, maintenance – become the "hows."

These four linked houses implicitly convey the voice of the customer through to manufacturing. A control knob setting of 3.6 gives an extruder speed of 100 rpm; this helps give a reproducible diameter for the weather-stripping bulb, which gives good sealing without excessive door-closing force. This feature aims to satisfy the customer's need for a dry, quiet car with an easy-to-close door.

None of this is simple. An elegant idea ultimately decays into process, and processes will be confounding as long as human beings are involved. But that is no excuse to hold back. If a technique like house of quality can help break down functional barriers and encourage teamwork, serious efforts to implement it will be many times rewarded.

What is also not simple is developing an organization capable of absorbing elegant ideas. The principal benefit of the house of quality is quality in-house. It gets people thinking in the right directions and thinking together. For most U.S. companies, this alone amounts to a quiet revolution.

Reprint 88307

Case Studies in Quality

GROWING CONCERNS

Instead of inspecting quality into its fasteners, the company learned to manufacture quality into the product.

How Velcro Got Hooked on Quality

by K. Theodor Krantz

The phone call came out of the blue one morning in August 1985. It was from our Detroit sales manager, who told me that General Motors was dropping us from its highest supplier quality rating to the next to lowest level, four (on a one to five scale). We had 90 days to set up and start a program of total quality control at our sole U.S. plant, in Manchester, New Hampshire, or face the loss of not only an important customer but also our most promising growth market.

The blow was especially hard because we were feeling pretty good about our quality control. About a year earlier we had sent our QC man to school to learn statistical process control and had brought in a consultant to talk to management for half a day and do some training with the quality control staff. Essentially we had done this the way you approach projects, like launching a new product or putting in a safety program in a certain area: you set up the project and choose a project team, then you delegate the work.

We had a reputation in the marketplace for products of the highest quality. True, a year earlier we had also made a toothless attempt to modernize Velcro's quality program. The effort had been delegated to the quality assurance manager, and it had ended stillborn. Nevertheless, the phone call came as a real shock.

To get an idea of what we had to do, we needed to talk to GM. So five

> **News from Detroit threatened the loss of a big customer.**

Velcro managers flew out to Detroit and met for a couple of hours with the buyer who had given the news to our sales manager, her boss, and two GM quality control people. They told us that our products – which included tape for binding car seat parts together and for binding fabric to the roof – were fine and we were meeting their delivery schedules. But they said our process was unacceptable: we were *inspecting* quality into the product, we were not *manufacturing* quality into the product.

They were dissatisfied with the fact that we were throwing away 5% or 6% or 8% of tape, depending on the product. They wanted quality maintained up and down the line to prevent such waste. And they said that to have the head QC person report to the head of manufacturing was unacceptable. They wanted him to report to me.

We flew back East, feeling thoroughly chastened. It was no consolation to realize that all three U.S. auto producers were feeling great pressure to upgrade their quality, cut costs, and reduce the number of suppliers. We were not alone in getting the heat from GM.

We had 90 days to clean up our act, and in that time we had heavy education requirements. We had to get all 500 Manchester employees through a quality course of varying degrees of intensity, we had to work up a fairly exhaustive quality manual, and we had to start the rudiments of a statistical process control program in the plant. I had a bias against consultants, but we were out of time. We hired a local group of consultants that specializes in quality control.

The consultants started an education program for hourly and salaried employees, with emphasis on problem-solving techniques. We installed statistical process control and started keeping detailed records. We set up a steering committee consisting of the heads of sales, finance, personnel, marketing, manufacturing, R&D, quality, and MIS, as well as me. It met – and still meets – mainly to monitor progress of improvement teams, relying in part on the data from the SPC system.

The consultants were very helpful in pushing and pushing us. We needed an outside agent to make us see we weren't progressing as well as we in management thought we were.

K. Theodor Krantz spent 20 years in manufacturing with American Standard, CertainTeed, and Saint Gobain, the French building products maker. In 1983, he became chief financial officer of Velcro Industries N.V., a Netherlands Antilles company. A year later, he became president of its subsidiary, Velcro USA.

I can recall the aggravation I felt in steering committee meetings when a consultant would relay concerns voiced by hourly employees that were surprises to us. The consultants knew more about problems in the plant than we did.

(There are many good reasons, by the way, why smaller businesses shy away from hiring consultants. For one thing, they may seem expensive in relation to the value they claim to offer. They may be unfamiliar with the business, so they take too much time getting up to speed. And they can disrupt the organization. But they are valuable for the expertise they bring, the cross-fertilization of ideas they spread, the benefits they can bring in a devil's advocate role, and – as evidenced in Velcro's case – the compression of time they can achieve.)

At the end of 90 days, we showed the General Motors people a PERT chart to indicate how far we had come.

Waste was canceling out production's value added.

They said we seemed to be on the right track. They would be back in six months, they said, to inspect us.

Naturally, we were pleased at this reprieve, but we really had only scratched the quality surface. It takes a lot more than 90 days for, say, the loom operator on the third shift really to believe in quality and act on that belief. It was a long time before we got quality instilled throughout the organization, before people were making changes for Velcro instead of for GM. But it has happened, as these figures on waste reduction as a percentage of total manufacturing expenses, by fiscal year, testify:

1987, 50% reduction from 1986
1988, 45% reduction from 1987

We got our house in relatively good order and soon were back in GM's good graces. We expected to move up to a two rating, but by that time the company had changed its supplier quality program to Targets of Excellence. Even as our quality improved,

the standard against which we are measured was moving as well. All the auto companies have upgraded their programs – Chrysler has its Quality Excellence and Pentastar programs, and Ford has Q1 and Total Quality Excellence – and what was acceptable three years ago is no longer good enough. As we participated in all three programs, we had concerns that we would face conflicting requirements.

Pervasive commitment

The General Motors people were right in acknowledging that our quality watch was OK. Actually, the strength of our quality program was a major reason why we had survived a severe decline in prices some years before – due to a capacity overhang in the tape industry caused by gearing up to meet demand for shoes – and the reason why we had gained a foothold with the Big Three automakers. But the General Motors people were also right in their claim that our program was faulty because we inspected quality into the product instead of manufacturing it.

We had QC people at a few points in the process, for example, at the point where the tape was rolled up for shipment to customers. Quality determination coming at the end of the production process had to be relayed back up the line, and such feedback is often incomplete, unreliable, and certainly untimely. Besides, we were throwing out a certain amount of product because it was defective or didn't meet specifications. We were wasting all the value we had added in the production process, like dyeing, coating, and slitting. Moreover, when we found a problem at the end of the process, it might have run for several days before being spotted. Needless to say, if you're on JIT, you can't handle problems in this manner and still meet schedules.

In Velcro's case, if a weave defect cropped up, by the time it was noticed thousands of yards of material might have been made. What a waste! We weren't being quality-effective, and we weren't being cost-effective either.

On top of that, there were internal problems. The consultants had

picked up on these in conversations with hourly workers in the plant. Changes had been coming thick and fast. First there had been heavy demand for footwear, which we couldn't meet. Then there had been a dramatic squeeze on margins from the import competition, making a shorter time frame in the manufacturing process necessary to improve service. The super emphasis on quality was one more change layered on top of the others.

At the start of the new drive to instill a quality ethic throughout the Velcro organization, I had envisioned a high degree of contagiousness for this "disease." The rationales were clear and simple. How could anyone argue with the well thought-out logic of W. Edwards Deming and other proponents?

☐ Do it right the first time.
☐ Employees are happier and more satisfied making good parts than making bad ones.
☐ Quality should be an important purchasing criterion.
☐ People must be trained properly.
☐ Teamwork is vital to success.

But even if there hadn't been bewilderment and discontent out in the plant, the major changes that management was seeking would have contributed enough "vaccine" to block progress of the disease. The solution was to involve everybody from the top down and show them that the organization's survival depended on improvement. The mes-

In the first year, we trimmed waste 50%; in the second year, 45% more.

sage from General Motors went a long way toward demonstrating that reality.

When you look at the manuals that both GM and Ford have put out in the last year, it's clear that their focus is not so much on the product but on continuous improvement. And the improvements sought are not just on the production line or the shop floor but also in the financial department, in the marketing department, in the sales department, and so on. When you're a supplier to these companies, they want to see your cost control programs and your cost reduction programs because they realize they're all tied together. That's how the Japanese have reached the position they enjoy in the automotive business.

It may be a cliché to say that you have to involve everybody in management, including the CEO – but it's true. Only if hourly and salaried employees see that top management is totally committed to quality will quality command a high priority with them. One way they see it is in meetings of the steering committee every two weeks to hear employees talk about their continuous improvement projects. I make it a point to sit in on every meeting, and if I'm out of town, the meeting is postponed. Moreover, I'm a member of one of the teams working on quality improvements.

So the rank and file has good reason to believe that a commitment to quality comes at the highest level. That's a big step, but it's only a step. An even bigger step is getting the hourly workers to understand that continuous improvement not only

Velcro made everyone responsible for quality – and halved its QC staff.

benefits them but also needs them. Opening up the channels and getting them to communicate is vital because, as Deming and Joseph M. Juran said, the people on the floor know the process better than their boss or anybody up the line. Many of the best ideas for building a better quality product are in their heads, waiting to spring out.

As our QC director is fond of pointing out, communication is the glue that holds the quality effort together. The company newsletter usually has a lead article featuring quality or

some of our continuous improvement teams. Weekly meetings with a handful of employees, me, and key staff members feature the same topic. Quarterly talks to all employees discuss progress in this area. The point is that the smaller size of a company makes communication to the organization easier (but still not easy!), and it is vital that the CEO exploit this advantage.

Administrative roadblocks

At the time we came under pressure from General Motors, we had 23 quality control people in the plant. (Now we have 12.) To the machine operators, quality was their responsibility – that is, someone else's. The quality control people were stationed at certain points, and they would inspect on a sample basis and say whether the particular run was good or bad. What was bad was thrown out. Nobody changed the process, there was no pressure on anybody to make a change. Many manufacturing people were extremely reluctant to take charge of quality; that was for the end of the line, when somebody else sorts out the errors.

To assume that the production employees were causing the waste would have been a mistake, and to beat on them about it without giving them the tools to deal with the problem would have been a bigger mistake. They would only have been afraid to report it. That's the fear element Deming talks about. The waste is going out in a Dumpster during the third shift, and management thinks it's running at minimal waste until it takes inventory. We knew we had to invest in operator training, in more attention paid to operators, in machine repair and redesign, and in measurement and reporting techniques that tracked results, focused on responsibilities, and established up-and-down communication.

Part of the reason for the lack of pressure for change was the supervisors. We'd hear comments like "My boss won't let me shut the machine down. We make junk on my shift, but he doesn't care. He just says we've got to get x yards of material out. I show him the material, and he says, 'Run it anyway.' " The supervisors

were a big barrier to making the operators responsible for quality.

The SPC system we installed went a long way toward pinpointing where in the production process we needed improvement. The charting mechanism of SPC also put pressure on the people on the line who had difficulty with the idea that quality and quan-

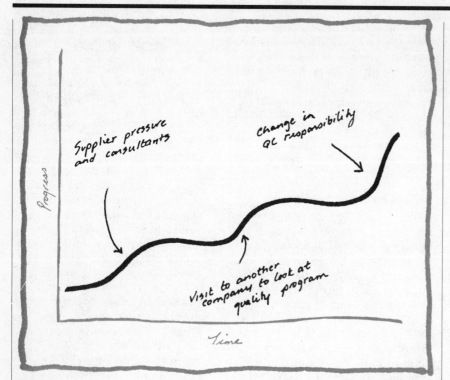

Supplier pressure and consultants

change in QC responsibility

Visit to another company to look at quality program

Progress

Time

tity aren't mutually exclusive expectations. The operators used to protest, "I see what you're saying, but I know that when I speed up the machine it makes more mistakes." Gradually we pulled the quality control people out of stations early in the process, then we pulled them out of points later in the process. The number of mistakes declined.

Ad hoc teams, which we call "continuous improvement" teams, not QC—to emphasize the fact that we can always do it better—were another powerful mechanism. We have had some 120 teams over the past three years and now have about 50. (Half of these are in the administrative area; I'll get to those in due course.) At first the teams were led by supervisors, but now hourly employees are sometimes at the head. We've had some spectacular savings in spare-parts consumption from a team led by an hourly employee.

To increase awareness, manufacturing started a material review board consisting of a rotating group of salaried and hourly people. Each day, they look at substandard material from each unit's production, discuss the causes of the defects, and decide on the disposition of the material. This helps to close the loop on awareness of problems, ensuring feedback to the responsible department and encouraging corrective action.

We haven't entirely overcome the difficulty of getting the supervisors to buy into the new way of doing things. The problem has deep roots. At Velcro, as at a lot of mature companies, many of these individuals were trained during an era when the boss was king. Supervisors who hark back to this era often view participatory problem solving as a threat to their authority. On top of that, hourly employees may grasp innovations like SPC more readily than their supervisors do, which tends to produce a defensive attitude regarding the procedures.[1]

Our approach has focused on building successful examples, especially among the younger supervisors who are often more receptive to new ideas. Once we had some "carriers" in these ranks, we could get a quality "infection" spread among many of their peers, whom they saw every day. One of these carriers is a supervisor in the coating area. Early in the days when we trained operators to be their own QC persons, he was one of them. After his promotion to supervisor, he came up with a procedure to reduce the sample size and improve the measurement of coating. His example helped people real-

ize that the new era wasn't all that threatening. He wasn't a graduate engineer; he had no more education than they did.

We try to get the supervisors to understand that the people who work under them are in many respects as good as they are. They have certain skills that can be used and built up. We try to show the supervisors that building up the rank and file doesn't weaken their position. On the contrary, when a production unit meets its goals or makes an improvement, they look good. Eventually, many supervisors come to accept the tenet of Deming and Juran's that the people on the floor know the process better than their boss or anybody up the line.

Even so, some supervisors never get the idea. We have had to terminate people who had been with Velcro a long time but who, after considerable effort on their part, hadn't gotten the message. They had become a roadblock for other people.

Spreading the gospel

In a manufacturing company, it's natural for everybody to think of the manufacturing side when the word "quality" comes up. Marketing and sales holler about the quality of the product, but the quality of what *they* do never comes under question. When they understand we're all in this together, we can get them involved.

That's why about half of our quality improvement teams are in departments other than manufacturing. One of the teams in finance is trying to improve the accuracy of the payroll. A tenet of quality is that you have inside customers as well as outside customers, and you have to satisfy both. When the paychecks are not correct, your inside customers get very upset, particularly the inevitable few who think the company is taking advantage of them. It's a very important morale issue. Moreover, responding to complaints and reissuing paychecks cost money.

Another team in finance has reduced the number of copies of pur-

1. For a valuable analysis, see Janice A. Klein, "Why Supervisors Resist Employee Involvement," HBR September-October 1984, p. 87.

chase orders from six to three and number of copies of invoices from six to four (the goal being two). The sales operation has a team evaluating the quality of its service. Sales managers are interviewing customers on the phone to ascertain how well we're meeting their needs. The cost of quality in sales is to an extent quantifiable, such as time spent in dealing with a customer's complaint and lost business.

Because the heads of the line and staff functions all serve on the steering committee, it is another way of spreading the word about quality throughout the organization. Function managers who cannot attend the biweekly meetings send deputies, thus ensuring continuity. Moreover,

What do you do about supervisors who hinder change?

employees giving presentations about their continuous improvement projects usually sit through the others scheduled for that day. So they get a wider view too.

It's not easy to maintain a high level of interest in a program like ours, so a periodic stimulus is needed. The accompanying chart sketches the benefits to Velcro of occasional shots in the arm. The first, of course, was the General Motors threat. The second was a benchmarking effort, a trip to a prospective customer that was also a GM supplier. This visit gave us new ideas and also confirmed the value of SPC charting. The third shot in the arm came when our QC head left and we hired a replacement. The former was good, and we feared a period of drifting. But under the new person, the program really took off. For yet another stimulus, we may bring the consultants back for a time.

There's no finish line

In the search for quality, there's no such thing as good enough; there's never a finish line. Moreover, the finish line sometimes seems farther away than ever. We have discovered that in our current year in the campaign against waste. When we realized significant waste reduction (as a proportion of total production expenses) in fiscal 1987 and 1988, we had skimmed the cream. Now it's much harder to make improvements, and we have been unable to lower the figure in our current year.

Still, Velcro has come a long way since the days when it enjoyed patent protection and didn't have to worry much about its customers. We gave the product to them when we wanted and in the form we wanted, because they couldn't get it anywhere else. Now Velcro is a customer-driven company where quality is in the eye of the customer. Whether a production run is acceptable is something the customer ultimately decides.

Sometimes quality is the customer's perception; it's not an absolute. It's all relative, depending on the customer's requirements. In the shoe industry, hook-and-loop on a pair of kid's sneakers isn't especially important since the goods barely last three months, if that. But with a $600 knee brace, the quality of the hook-and-loop closure is very important.

For the textile customer, appearances are important, whereas in the medical business the concern is cleanliness. The automakers want durability, reliability, and capability. With government, the specifications are all-important. Car manufacturers want uniform performance, but they wouldn't care if the weave on the Velcro tape's backing is not uniform. But on the knee brace, where appearance is important, the weave has to be right.

Since quality is relative, the supplier must be sensitive to the customer's particular requirements. They may involve a varying mix of dimension, performance, cosmetic characteristics, or chemical elements. Here is an essential element of quality that doesn't necessarily show up in the SPC charts. But it's part of the goal of customer satisfaction that creates barriers to competition.

Reprint 89508

The Case of the Quality Crusader

Segal Electric says it cares about quality. So why does it ship defective fans?

by FRANK S. LEONARD

George Mansfield poured himself another cup of coffee and walked down the hall to Pete Jameson's office. He'd spent half the night agonizing over how to broach a problem with Pete, the general manager who had hired him a month ago to run Segal Electric's quality operation as director of quality assurance.

George had always thought that anything worth making was worth making right. That's one reason he advocated generous warranty programs. He knew that if operations were running as they should, the company wouldn't have to worry about warranty costs because there wouldn't be any. His high standards had served him well at his last job, where he was perceived as a real comer and had been promoted several times.

Segal was once the top manufacturer of high-quality electrical products, but for a combination of reasons its reputation had fallen steadily over the last few years. George welcomed the chance to help turn the company around. The hefty salary and direct reporting relationship to Pete Jameson had persuaded him that Segal was serious about regaining its position in the market. George was given much of the responsibility to help reach that goal.

"Come on in, George, and have a seat," Pete began. "So you've run into some trouble in the plant already. I told you this job would be tough. What's the problem?"

"There was a mix-up in the plant while I was gone last week. I think it might happen again, so I want to discuss it with you now."

"Of course, go ahead."

"Last week when I was at that quality seminar in Milwaukee, Gene Davis was overseeing the quality control team for me. Early in the week, they kept running into trouble with the 5051 fans. The end play of the fan blades on the shaft wasn't right. They had to reject five pallets of fans.

"Apparently there was a big rush order, so some people were pulled off other lines to assemble the fans. But nobody told them all the specs. So they were loading up as many thrust washers as they could fit on the shaft, and instead of getting ten-thousandths clearance, they got zero. There was no end play at all.

"Following quality procedures, Gene red-tagged them and put them aside for rework, but then Sharon Morse got involved. She decided it would take too long to fix the fans, so she plugged the darn things in and ran them full speed to burn the washers down to the right clearance. Now, you and I both know that doing that pushes the spec strength of those washers."

"Well, in theory, yes. But I don't know how much difference it really makes. Still, I get your point, George. That's not the way it ought to be."

George: "Doesn't QC have the last word on what gets shipped?"

"No sir, as far as I'm concerned it isn't. I guess Gene tried to talk to Sharon about it. He knows how I feel about quality, and he was trying to do what I would expect. But Sharon took charge of the situation. As production superintendent she outranks Gene, and he finally deferred to her. Well, don't you know, those fans were shipped last Thursday."

Frank S. Leonard is an independent executive educator and consultant. He specializes in the strategic use of quality and manufacturing, particularly for large multinational corporations.

"Have you discussed this with Sharon?" Pete asked.

"No, I haven't. Gene just told me about it yesterday. I'm perfectly willing to talk to her, but I want to make sure I'm right first. Doesn't my department have the last word about what gets out the door?"

"George, we're like every other company. We've got production quotas, we've got budgets, we've got marketing plans. Sure, we want to make a good product, but we don't want to lose our shirts doing it. Sounds to me like you and Sharon will have to work it out. I think you can handle it."

Sharon: "It's on-time delivery that gets us business."

George was disappointed by Pete's response. After all, in the job interview Pete assured him that Segal was serious about the renewed emphasis on quality. But it didn't seem as if the idea had filtered down to the factory floor—where priorities are really set.

The people in the plant were very loyal to Sharon. According to Gene Davis, they had often helped her patch up rejected products and rush them onto the shipping dock to help make quotas. George was anxious to confront her. Later that day when he saw her on the plant floor, he made a point of starting a conversation. He didn't want to come on too strong, but he wanted to let her know he wasn't happy about what had happened last week and that he didn't want it to happen again.

"Say, how did things go when I was gone?" he asked.

"It was touch and go for a while there, but we managed to fill a couple of big, important orders."

"I understand you were short of assemblers."

"We're always short of assemblers. But you're right. Last week absenteeism was especially high, and we had to do a lot of juggling to meet our quotas. You know about the special promotions?"

"I've heard of something, but no details."

"Sales is running a special seasonal promotion for both trade and retail. It's in response to a new product that some competitor is coming out with. We're trying to preempt that market share."

"Gene tells me we ran into some trouble with the 5051s."

"Yes, sales was leaning all over us for them. All I kept hearing all week was 'Delivery! Delivery!'"

"But some of the products were rejected?"

"Well, the thrust washers were tight. But you know, if you run them tight for a little while, it frees them up just the right amount. You should have seen this place with all those fans running. It was quite a sight."

"But doesn't running them tight like that burn the washers and stress the metal?"

"Well, I'm not so sure about that."

George was annoyed by Sharon's skepticism, especially since he felt sure of himself on technical matters. After all, he had been trained as an engineer. Sharon had started as a production worker and come up through the ranks. Despite the fact that George had been at Segal for only a month, he was sure Sharon didn't understand the product as well as he did.

"I don't want to make a big deal of something that's already happened, but it's not such a good idea to burn the washers into spec. We should be putting the product together correctly in the first place."

"Oh boy, you won't get any disagreement from me there. I had some people who started work on Monday and were on their own assembling parts that afternoon. I barely had time to show them where to hang their coats."

"Don't new employees get any training? At my old company everyone was cross-trained so they could do at least a couple of different tasks. That way, people could be moved from one job to another."

"Do you know how big our training department is? Do you know what our turnover is? We're lucky if we have *one* person who knows how to do the job. But you know, some-times the problem is with the parts we get—not so much the ones we get from our shop but the ones we get through purchasing. I don't know whether they're mixing up part numbers or what, but those washers aren't all the same."

"What makes you say that?"

"When you put them on the fan shaft, you can tell. One fan has five washers and twenty-thousandths clearance. Another fan has five washers and five-thousandths clearance. If you ask me, they're getting the cheapest washers they can find."

"Of course, if it's just a slight variation, the assemblers can adjust for it when they're putting the fan together," George conceded.

"Are you kidding? We can't pull out a micrometer every time we throw a few thrust washers onto a shaft. We'd never get our orders out. The bill of material and assembly specs for a 5051 fan call for five thrust washers, so that's what we use. Every once in a while a worker forgets that. Like last week, when we were in such a rush. Some of the people hadn't worked on that line for months, so they were putting on five, six, seven of them—however many they could fit on the shaft. But normally it's five washers. Look George, if you want to measure those washers, I'd be real happy. I'm all for anything that will help me meet my objectives."

Pete: "Sure we want to make a good product, but we don't want to lose our shirts."

"I'm not interested in measuring washers, but we should be checking that clearance before the fans get to final test. It would take just one second to make the adjustment during assembly, and you know how long it takes to pull the thing apart and re-make it."

"It's not that I don't have a mind of my own, but I follow the process orders that come from engineering.

That's my job. I'm not going to arbitrarily change the procedure. Besides, five washers is usually right, so there's really not a problem. Do you know we ship 95% of our orders on time? We're pretty proud of that record. Sales tells me that it's our delivery performance that gets us business. I don't want to mess with that."

The conversation with Sharon ended on a friendly note, but George knew he hadn't won her over. He walked back to his office to decide what to do next.

WHAT'S THE WAY OUT?

We asked the following business leaders—people who actually have to deal with such problems—what went wrong at Segal and how they would solve the dilemma. Here are their responses.

WILLIAM A. GOLOMSKI *is president of W.A. Golomski & Associates, a Chicago-based international quality and productivity consulting group.*

Segal doesn't understand quality.

Segal is in trouble. The company does not understand total quality control (the use of statistical concepts at all levels), and it lacks sound principles of leadership and team effort. Moreover, purchasing, sales, engineering, human resources, and general management simply do not understand their roles in quality. The situation in the case is common in companies whose CEO has a legal or accounting background. It occurs less often when the CEO has marketing or technical experience.

No one at Segal understands variability. Statistical thinking is absent in both the production and quality functions. Practical statistical concepts are simple enough for those without a college degree to understand, yet no one at Segal has been trained to use them.

George Mansfield is not a leader and lacks good judgment. There is no reason for him to have run off to a seminar after being on the job such a short time. And Mansfield failed to brief Gene Davis adequately, or to inform the rest of the organization that Davis was in charge while Mansfield was gone.

Also, Mansfield does not think as broadly as he should. Warranty programs are usually designed by marketing with input from the reliability side of quality assurance. So while Mansfield may have an opinion on warranty strategy, his ideas are only one part of it. Often warranty claims are related not to operations but to design. (I know of a situation where 60% of the problems in the manufacture and use of mechanical and electrical products were related to design.)

Careful selection and training of employees—one of the best ways to prevent product problems—is not a priority at Segal. Senior management seems to use detection as the primary mode of control, but then only if it doesn't get in the way of schedule.

This is not a world-class quality company and it won't survive.

Segal should worry about customers — not quotas.

RAY J. ROGAL *is director of quality for the Ford Motor Company in Dearborn, Michigan. He implements quality policies and philosophies at Ford worldwide.*

Segal Electric's loss of market share was at least partially due to its preoccupation with internal performance indicators as the absolute surrogate for the customer. This kept it from focusing on customers' real needs in terms of quality, cost, and value.

Apparently Pete Jameson believes that internal performance targets are foremost. When an issue arises involving the quality of products provided to customers, he immediately refers to quotas, budgets, and marketing plans. While these are important concerns, Pete's failure to mention quality tells a lot about the culture at Segal and the way it views quality. Pete is all for quality until budgets and quotas are at risk. It's no wonder that production personnel don't emphasize quality when push comes to shove.

An enlightened view of Segal Electric's focus — a quality philosophy — might be, "To provide customers with products and services that meet customers' expectations at a price that represents value." A quality philosophy in itself does not change the way a company is managed. The quality philosophy must be an integral part of the culture and environment — and of how employees think about their jobs. It must be the basis on which every business decision is made. When consistently implemented throughout the company, the philosophy puts issues of budget and schedule performance into perspective.

Such a philosophy would immediately raise a key question that went unasked in this case: "How will this product condition (motor and play and subsequent 'run-in') affect our customers?" This seems an incredible omission for a company that's trying to regain market share. While the customer's interest in on-time delivery apparently is well-known throughout the company, other matters of customer interest, like proper function and durability, seem to have escaped management's attention.

To refocus its definition of quality, Segal should consider these questions:
■ How do we communicate customer needs and expectations to our production people?
■ How should we educate and train employees in quality?
■ What effect does the concept of variability have on incoming materials and the finished product in terms of consistency and customer satisfaction?
■ How do we control the *quality of our products* (not quality control)?
■ How do we continuously improve in order to better satisfy both current and potential customers?

Segal needs better communication among departments.

KEITH LARSON *is general manager of the Eastern Operation for Barry Controls, a Watertown, Massachusetts manufacturer of shock vibration- and noise-control devices.*

The case depicts a problem that is all too familiar in industry today: lack of communication. This issue surfaces first in the way Pete Jameson communicates his commitment to quality down through the ranks. It shows up again among Jameson's subordinates (namely, quality assurance, engineering, and manufacturing) when problems arise on the factory floor.

Of course, words alone are not enough. As general manager, Jameson should be well aware that his actions must complement his verbal commitments. His response to George Mansfield about the defective fans illustrates his lack of *total* commitment to quality, and his subordinates will use that behavior as a guide in making their decisions.

The lack of communication between departments is quite common in manufacturing organizations. There is always pressure, for multitudes of reasons, to push product shipments, and this can usually be done in a rational manner while maintaining high quality. When problems arise, engineering, quality, and manufacturing should join forces immediately and decide what to do. The solution may include a bill of material change, increased inspection points, or process changes.

In most cases, engineering should take the lead in deciding corrective action and/or design refinements, and these actions should be implemented in a timely fashion. Doing so requires good, strong communication among departments. Communication is paramount to building a quality product and maintaining shipping schedules.

At Segal, the process was changed without engineering even knowing about it, let alone approving it. Mansfield has every right to be disturbed by the way the situation was handled. He also has reason to be disappointed in Jameson's response.

Jameson's desire for high quality has not been transmitted properly throughout the organization. All decisions made within the organization must reflect the importance of quality. Our actions usually dictate where our true commitments lie.

Pete and George have to prove their commitment through actions — not words.

MICHAEL T. COWHIG *is vice president of manufacturing at Gillette's Personal Care Group, where he develops manufacturing strategy and product sourcing plans for both Europe and North America.*

Segal Electric's attempt to restore its reputation for high quality and regain market share requires more than the addition of a highly paid quality assurance director. Hiring George Mansfield is a good first step, but unless everyone at Segal, from Pete Jameson on down, changes the approach to quality improvement, the effort will fail.

Jameson's policy of "quota first, quality second" sets the tone for the entire organization. Sharon Morse is getting a clear signal from the top that meeting quota is her primary objective. This damaging message must be changed by setting and measuring quotas on the basis of defect-free units only.

But that one change is not enough. In fact, without corresponding changes in training, job definition, and purchasing policy, new quota standards will only increase conflict and frustration.

Other changes should include training assemblers. They should be taught about product and about the fundamentals of quality. Also, purchasing should include the cost of rework and late delivery in the total cost of purchased parts. After all, if the washers had met the specs, the fans would have met both quality and delivery goals.

Changing company practices will eventually enable Segal to use improved product quality as a strategic tool. But things are not going to improve overnight. The old quota system has no doubt been in place for some time now, and neither George Mansfield nor Pete Jameson can expect to legislate an attitude change. They will have to convince the organization that top management is committed to improving product quality through actions rather than words.

Jameson should view the problem with the 5051 fans as an opportunity to demonstrate his commitment to quality improvement by taking action. And Mansfield, too, should see the 5051s as an opportunity. He should convince Jameson that improved quality does not require a trade-off in cost or customer service. Indeed, given the expenses associated with warranties, rework, and lead time, quality improvement actually lowers costs and improves customer service.

At the Personal Care Group of the Gillette Company, production operators and inspectors make store checks to see firsthand how their efforts affect the product on the shelf. Quality is clearly a production department responsibility. On many of the lines, the operators and mechanics inspect the items. This practice cuts the time between quality checks and process adjustments and results in lower scrap and reject rates.

Plant purchasing and quality control teams coordinate an aggressive "supplier quality process," which includes vendor certification and recognition programs. Other teams composed of R&D and manufacturing representatives work on new product specifications to build quality into product and process designs before launch.

The key to making all these efforts work, however, is top management commitment. At Gillette — as at Segal — quality improvement efforts succeed only when the direction is clear, consistent, and supported by the actions of top management.

Reprint 88308

Service Quality

*No-quibble guarantees are self-fulfilling —
they promise quality and produce it.*

The Power of Unconditional Service Guarantees

by CHRISTOPHER W.L. HART

When you buy a car, a camera, or a toaster oven, you receive a warranty, a guarantee that the product will work. How often do you receive a warranty for auto repair, wedding photography, or a catered dinner? Virtually never. Yet it is here, in buying services, that the assurance of a guarantee would presumably count most.

Many business executives believe that, by definition, services simply can't be guaranteed. Services are generally delivered by human beings, who are known to be less predictable than machines, and they are usually produced at the same time they are consumed. It is one thing to guarantee a camera, which

> **Al Burger *started* with an unconditional guarantee and built his company around it.**

can be inspected before a customer sets eyes on it and which can be returned to the factory for repairs. But how can you preinspect a car tune-up or send an unsuccessful legal argument or bad haircut back for repair? Obviously you can't.

But that doesn't mean customer satisfaction can't be guaranteed. Consider the guarantee offered by

"Bugs" Burger Bug Killers (BBBK), a Miami-based pest-extermination company that is owned by S.C. Johnson & Son.

Most of BBBK's competitors claim that they will reduce pests to "acceptable levels"; BBBK promises to eliminate them entirely. Its service guarantee to hotel and restaurant clients promises:

■ You don't owe one penny until all pests on your premises have been eradicated.

■ If you are ever dissatisfied with BBBK's service, you will receive a refund for up to 12 months of the company's services — plus fees for another exterminator of your choice for the next year.

■ If a guest spots a pest on your premises, BBBK will pay for the guest's meal or room, send a letter of apology, and pay for a future meal or stay.

■ If your facility is closed down due to the presence of roaches or rodents, BBBK will pay any fines, as well as all lost profits, *plus* $5,000.

In short, BBBK says, "If we don't satisfy you 100%, we don't take your money."

Christopher W.L. Hart is an assistant professor at the Harvard Business School, where he teaches a course on service management. As a researcher and consultant, he helps companies design and implement service-guarantee and quality-improvement programs.

How successful is this guarantee? The company, which operates throughout the United States, charges up to ten times more than its competitors and yet has a disproportionately high market share in its operating areas. Its service quality is so outstanding that the company rarely needs to make good on its guarantee (in 1986 it paid out only $120,000 on sales of $33 million—just enough to prove that its promises aren't empty ones).

A main reason that the "Bugs" Burger guarantee is a strong model for the service industry is that its founder, Al Burger, began with the concept of the unconditional guarantee and worked backward, designing his entire organization to support the no-pests guarantee—in short, he started with a vision of error-free service. In this article, I will explain why the service guarantee can help your organization institutionalize superlative performance.

What a Good Service Guarantee Is

Would you be willing to offer a guarantee of 100% customer satisfaction—to pay your dissatisfied customer to use a competitor's service, for example? Or do you believe that promising error-free service is a crazy idea?

Not only is it not crazy, but *committing* to error-free service can help force a company to *provide* it. It's a little like skiing. You've got to lean over your skis as you go down the hill, as if willing yourself to fall. But if you edge properly, you don't fall or plunge wildly; you gain control while you pick up speed.

Similarly, a strong service guarantee that puts the customer first doesn't necessarily lead to chaos and failure. If designed and implemented properly, it enables you to get control over your organization—with clear goals and an information network that gives you the data you need to improve performance. BBBK and other service companies show that a service guarantee is not only possible—it's a boon to performance and profits and can be a vehicle to market dominance.

Most existing service guarantees don't really do the job: they are limited in scope and difficult to use. Lufthansa guarantees that its customers will make their connecting flights *if* there are no delays due to weather or air-traffic control problems. Yet these two factors cause fully 95% of all flight delays. Bank of America will refund up to six months of checking-account fees if a customer is dissatisfied with any aspect of its checking-account service. However, the customer must close the account to collect the modest $5 or $6 per month fee. This guarantee won't win

any prizes for fostering repeat business—a primary objective of a good guarantee.

A service guarantee loses power in direct proportion to the number of conditions it contains. How effective is a restaurant's guarantee of prompt service *except* when it's busy? A housing inspector's guarantee to identify all potential problems in a house *except for* those not readily apparent? Squaw Valley in California guarantees "your money back" to any skier who has to wait more than ten minutes in a lift line. But it's not that easy: the skier must first pay $1 and register at the lodge as a beginner, intermediate, or expert; the guarantee is operative only if *all* lifts at the skier's skill level exceed the ten minutes in any half-hour period; and skiers must check with a "ski hostess" at the end of the day to "win" a refund. A Squaw Valley spokesperson said the resort had made just one payout under the guarantee in a year and a half. No wonder!

What is a good service guarantee? It is (1) unconditional, (2) easy to understand and communicate, (3) meaningful, (4) easy (and painless) to invoke, and (5) easy and quick to collect on.

Unconditional. The best service guarantee promises customer satisfaction unconditionally, without exceptions. Like that of L.L. Bean, the Freeport, Maine retail store and mail-order house: "100% satisfaction in every way...." An L.L. Bean customer can return a product at any time and get, at his or her option, a replacement, a refund, or a credit. Reputedly, if a customer returns a pair of L.L. Bean boots after ten years, the company will replace them with new boots and no questions. Talk about customer assurance!

Customers shouldn't need a lawyer to explain the "ifs, ands, and buts" of a guarantee—because ideally there shouldn't be any conditions; a customer is either satisfied or not.

If a company cannot guarantee all elements of its service unconditionally, it should unconditionally guarantee the elements that it can control. Lufthansa cannot promise on-time arrival, for example, but it could guarantee that passengers will be satisfied with its airport waiting areas, its service on the ground and in the air, and its food quality—or simply guarantee overall satisfaction.

Easy to Understand and Communicate. A guarantee should be written in simple, concise language that pinpoints the promise. Customers then know precisely what they can expect and employees know precisely what's expected of them. "Five-minute" lunch service, rather than "prompt" service, creates clear expectations, as does "no pests," rather than "pest control."

Meaningful. A good service guarantee is meaningful in two respects. First, it guarantees those aspects of your service that are important to your customers. It may be speedy delivery. Bennigan's, a restaurant chain, promises 15-minute service (or you get a free meal) at lunch, when many customers are in a hurry to get back to the office, but not at dinner, when fast service is not considered a priority to most patrons.

In other cases, price may be the most important element, especially with relatively undifferentiated commodities like rental cars or commercial air travel. By promising the lowest prices in town, stereo shops assuage customers' fears that if they don't go to every outlet in the area they'll pay more than they ought to.

 ## L.L. Bean will replace its boots— even after ten years' use.

Second, a good guarantee is meaningful financially; it calls for a significant payout when the promise is not kept. What should it be—a full refund? An offer of free service the next time? A trip to Monte Carlo? The answer depends on factors like the cost of the service, the seriousness of the failure, and customers' perception of what's fair. A money-back payout should be large enough to give customers an incentive to invoke the guarantee if dissatisfied. The adage "Let the punishment fit the crime" is an appropriate guide. At one point, Domino's Pizza (which is based in Ann Arbor, Michigan but operates worldwide) promised "delivery within 30 minutes or the

pizza is free." Management found that customers considered this too generous; they felt uncomfortable accepting a free pizza for a mere 5- or 15-minute delay and didn't always take advantage of the guarantee. Consequently, Domino's adjusted its guarantee to "delivery within 30 minutes or $3 off," and customers appear to consider this commitment reasonable.

Easy to Invoke. A customer who is already dissatisfied should not have to jump through hoops to invoke a guarantee; the dissatisfaction is only exacerbated when the customer has to talk to three different people, fill out five forms, go to a different location, make two telephone calls, send in written proof of purchase with a full description of the events, wait for a written reply, go somewhere else to see someone to verify all the preceding facts, and so on.

Traveler's Advantage—a division of CUC International—has, in principle, a great idea: to guarantee the lowest price on the accommodations it books. But to invoke the guarantee, customers must prove the lower competing price by booking with another agency. That's unpleasant work. Cititravel, a subsidiary of Citicorp, has a better approach. A customer who knows of a lower price can call a toll-free number and speak with an agent, as I did recently. The agent told me that if I didn't have proof of the lower fare, she'd check competing airfares on her computer screen. If the lower fare was there, I'd get that price. If not, she would call the competing airline. If the price was confirmed, she said, "We'll refund your money so fast, you won't believe it—because we want you to be our customer." That's the right attitude if you're offering a guarantee.

Similarly, customers should not be made to feel guilty about invoking the guarantee—no questioning, no raised eyebrows, or "Why me, Lord?" looks. A company should encourage unhappy customers to invoke its guarantee, not put up roadblocks to keep them from speaking up.

Easy to Collect. Customers shouldn't have to work hard to collect a payout, either. The procedure should be easy and, equally important, quick—on the spot, if possible. Dissatisfaction with a Manpower temporary worker, for instance, results in an immediate credit to your bill.

What you should *not* do in your guarantee: don't promise something your customers already expect; don't shroud a guarantee in so many conditions that it loses its point; and don't offer a guarantee so mild

1. See British Airways study cited in Karl Albrecht and Ron Zemke, *Service America!* (Homewood, Ill.: Dow Jones-Irwin, 1985), pp. 33-34.

that it is never invoked. A guarantee that is essentially risk free to the company will be of little or no value to your customers—and may be a joke to your employees.

Why a Service Guarantee Works

A guarantee is a powerful tool—both for marketing service quality and for achieving it—for five reasons. First, it pushes the entire company to focus on customers' definition of good service—not on executives' assumptions. Second, it sets clear performance standards, which boost employee performance and morale. Third, it generates reliable data (through payouts) when performance is poor. Fourth, it forces an organization to examine its entire service-delivery system for possible failure points. Last, it builds customer loyalty, sales, and market share.

A guarantee forces you to focus on customers. Knowing what customers want is the sine qua non in offering a service guarantee. A company has to identify its target customers' expectations about the elements of the service and the importance they attach to each. Lacking this knowledge of customer needs, a company that wants to guarantee its service may very well guarantee the wrong things.

British Airways conducted a market study and found that its passengers judge its customer services on four dimensions:[1]

1. Care and concern (employees' friendliness, courtesy, and warmth).

2. Initiative (employees' ability and willingness to jockey the system on the customer's behalf).

3. Problem solving (figuring out solutions to customer problems, whether unusual or routine—like multiflight airline tickets).

4. Recovery (going the extra yard, when things go wrong, to handle a particular problem—which includes the simple but often overlooked step of delivering an apology).

British Airways managers confessed that they hadn't even thought about the second and fourth categories. Worse, they realized that if *they* hadn't understood these important dimensions of customer service, how much thought could their employees be giving to them?

A guarantee sets clear standards. A specific, unambiguous service guarantee sets standards for your organization. It tells employees what the company stands for. BBBK stands for pest elimination, not pest control; Federal Express stands for "absolutely, positively by 10:30 A.M.," not "sometime tomorrow, probably." And it forces the company to define each

employee's role and responsibilities in delivering the service. Salespeople, for example, know precisely what their companies can deliver and can represent that accurately—the opposite of the common situation in which salespeople promise the moon and customers get only dirt.

This clarity and sense of identity have the added advantage of creating employee team spirit and pride. Mitchell Fromstein, president and CEO of Manpower, says, "At one point, we wondered what the marketing impact would be if we dropped our guarantee. We figured that our accounts were well aware of the guarantee and that it might not have much marketing power anymore. Our employees' reaction was fierce—and it had a lot less to do with marketing than with the pride they take in their work. They said, 'The guarantee is proof that we're a great company. We're willing to tell our customers that if they don't like our service for any reason, it's our fault, not

A service guarantee is valued when a customer's ego is on the line.

theirs, and we'll make it right.' I realized then that the guarantee is far more than a simple piece of paper that puts customers at ease. It really sets the tone, externally and, perhaps more important, internally, for our commitment to our customers and workers."

A payout that creates financial pain when errors occur is also a powerful statement, to employees and customers alike, that management demands customer satisfaction. A significant payout ensures that both middle and upper management will take the service guarantee seriously; it provides a strong incentive to take every step necessary to deliver. A manager who must bear the full cost of mistakes has ample incentive to figure out how to prevent them from happening.

A guarantee generates feedback. A guarantee creates the goal; it defines what you must do to satisfy your customers. Next, you need to know when you go wrong. A guarantee forces you to create a system for discovering errors—which the Japanese call "golden nuggets" because they're opportunities to learn.

Arguably the greatest ailment afflicting service companies is a lack of decent systems for generating and acting on customer data. Dissatisfied service customers have little incentive to complain on their own, far less so than unhappy product owners do. Many elements of a service are intangible, so consumers who receive poor service are often left with no evidence to support their complaints. (The customer believes the waiter was rude; perhaps the waiter will deny it.) Second, without the equivalent of a product warranty, customers don't know their rights. (Is 15 minutes too long to wait for a restaurant meal? 30 minutes?) Third, there is often no one to complain to—at least no one who looks capable of

> ## Without a guarantee, customers won't complain. Or come back.

solving the problem. Often, complaining directly to the person who is rendering poor service will only make things worse.

Customer comment cards have traditionally been the most common method of gathering customer feedback on a company's operations, but they, too, are inadequate for collecting valid, reliable error data. In the first place, they are an impersonal form of communication and are usually short (to maximize the response rate). Why bother, people think, to cram the details of a bad experience onto a printed survey form with a handful of "excellent—good—fair" check-off boxes? Few aggrieved customers believe that com-

pleting a comment card will resolve their problems. Therefore, only a few customers—usually the most satisfied and dissatisfied—provide feedback through such forms, and fewer still provide meaningful feedback. As a broad gauge of customer sentiment, cards and surveys are useful, but for specific information about customer problems and operational weaknesses, they simply don't fill the bill.

Service companies thus have a hard time collecting error data. Less information on mistakes means fewer opportunities to improve, ultimately resulting in more service errors and more customer dissatisfaction—a cycle that management is often unaware of. A guarantee attacks this malady by giving consumers an incentive and a vehicle for bringing their grievances to management's attention.

Manpower uses its guarantee to glean error data in addition to allaying customer worries about using an unknown quantity (the temporary worker). Every customer who employs a Manpower temporary worker is called the first day of a one-day assignment or the second day of a longer assignment to check on the worker's performance. A dissatisfied customer doesn't pay—period. (Manpower pays the worker, however; it assumes complete responsibility for the quality of its service.) The company uses its error data to improve both its work force and its proprietary skills-testing software and skills data base—major elements in its ability to match worker skills to customer requirements. The information Manpower obtains before and after hiring enables it to offer its guarantee with confidence.

A guarantee forces you to understand why you fail. In developing a guarantee, managers must ask questions like these: What failure points exist in the system? If failure points can be identified, can their origins be traced—and overcome? A company that wants to promise timely service delivery, for example, must first understand its operation's capability and the factors limiting that capability. Many service executives, lacking understanding of such basic issues as system throughput time, capacity, and process flow, tend to blame workers, customers, or anything *but* the service-delivery process.

Even if workers *are* a problem, managers can do several things to "fix" the organization so that it can support a guarantee—such as design better recruiting, hiring, and training processes. The pest-control industry has historically suffered from unmotivated personnel and high turnover. Al Burger overcame the status quo by offering higher than average pay (attracting a higher caliber of job candidate), using a vigorous screening program (making those hired feel like members of a select group), training all workers for six months, and keeping them motivated by giving them a great deal of autonomy and lots of recognition.

Some managers may be unwilling to pay for an internal service-delivery capability that is above the industry average. Fine. They will never have better than average organizations, either, and they will

A guarantee uncovers errors— and opportunities to learn.

therefore never be able to develop the kind of competitive advantage that flows from a good service guarantee.

A guarantee builds marketing muscle. Perhaps the most obvious reason for offering a strong service guarantee is its ability to boost marketing: it encourages consumers to buy a service by reducing the risk of the purchase decision, and it generates more sales to existing customers by enhancing loyalty. In the last ten years, Manpower's revenues have mushroomed from $400 million to $4 billion. That's marketing impact.

Keeping most of your customers and getting positive word of mouth, though desirable in any business, are particularly important for service companies. The net present value of sales forgone from lost customers—in other words, the cost of customer dissatisfaction—is enormous. In this respect, it's fair to say that many service companies' biggest competitors are themselves. They frequently spend huge amounts of money to attract new customers without ever figuring out how to provide the consistent service they promise to their existing customers. If customers aren't satisfied, the marketing money has been poured down the drain and may even engender further ill will. (See the insert, "Maximizing Marketing Impact.")

A guarantee will only work, of course, if you start with commitment to the customer. If your aim is to minimize the guarantee's impact on your organization but to maximize its marketing punch, you won't succeed. In the long run, you will nullify the guarantee's potential impact on customers, and your marketing dollars will go down the drain.

Phil Bressler, owner of 18 Domino's Pizza franchises in the Baltimore, Maryland area, demonstrates the right commitment to customers. He got upset the time his company recorded its highest monthly earnings ever because, he correctly figured, the profits had come from money that should have been paid out on the Domino's guarantee of "delivery within 30 minutes or $3 off." Bressler's unit managers,

who have bottom-line responsibility, had pumped up their short-term profits by failing to honor the guarantee consistently. Bressler is convinced that money spent on guarantees is an investment in customer satisfaction and loyalty. He also recognizes that the guarantee is the best way to identify weak operations, and that guarantees not acted on are data not collected.

Compare Bressler's attitude with that of an owner of several nationally franchised motels. *His* guarantee promises that the company will do "everything possible" to remedy a customer's problem; if the problem cannot be resolved, the customer stays for free. He brags that he's paid, on average, refunds for only two room guarantees per motel per year—a minuscule percentage of room sales. "If my managers are doing their jobs, I don't have to pay out for the guarantee," he says. "If I do have to pay out, my managers are not doing their jobs, and I get rid of them."

Clearly, more than two guests of *any* hotel are likely to be dissatisfied over the course of a year. By seeking to limit payouts rather than hear complaints, this owner is undoubtedly blowing countless opportunities to create loyal customers out of disgruntled ones. He is also losing rich information about which of his motels need improvement and why, information that can most easily be obtained from customer complaints. You have to wonder why he offers a guarantee at all, since he completely misses the point.

Why You May Need a Guarantee Even If You Don't Think So

Of course, guarantees may not be effective or practicable for all service firms. Four Seasons Hotels, for example, could probably not get much marketing or operational mileage from a guarantee. With its strong internal vision of absolute customer satisfaction, the company has developed an outstanding service-delivery system and a reputation to match. Thus it already has an implicit guarantee. To advertise the obvious would produce little gain and might actually be perceived as incongruent with the company's prestigious image.

A crucial element in Four Seasons's service strategy is instilling in all employees a mission of absolute customer satisfaction and empowering them to do whatever is necessary if customer problems do occur. For example, Four Seasons's Washington hotel was once asked by the State Department to make room for a foreign dignitary. Already booked to capacity, Four Seasons had to tell four other customers

Maximizing Marketing Impact

The odds of gaining powerful marketing impact from a service guarantee are in your favor when one or more of the following conditions exist:

The price of the service is high. A bad shoe shine? No big deal. A botched $1,000 car repair is a different story; a guarantee is more effective here.

The customer's ego is on the line. Who wants to be seen after getting a bad haircut?

The customer's expertise with the service is low. When in doubt about a service, a customer will choose one that's covered by a guarantee over those that are not.

The negative consequences of service failure are high. As consumers' expected aggravation, expense, and time lost due to service failure increase, a guarantee gains power. Your computer went down? A computer-repair service with guaranteed response and repair times would be the most logical company to call.

The industry has a bad image for service quality—like pest-control services, security guards, or home repair. A guard company that guarantees to have its posts filled by qualified people would automatically rank high on a list of prospective vendors.

The company depends on frequent customer repurchases. Can it exist on a never-ending stream of new triers (like small service businesses in large markets), or does it have to deal with a finite market? If the market is finite, how close is market saturation? The smaller the size of the potential market of new triers, the more attention management should pay to increasing the loyalty and repurchase rate of existing customers—objectives that a good service guarantee will serve.

The company's business is affected deeply by word of mouth (both positive and negative). Consultants, stockbrokers, restaurants, and resorts are all good examples of services where there are strong incentives to minimize the extent of customer dissatisfaction—and hence, negative word of mouth.

with reservations that they could not be accommodated. However, the hotel immediately found rooms for them at another first-class hotel, while assuring them they would remain registered at the Four Seasons (so that any messages they received would be taken and sent to the other hotel). When rooms became available, the customers were driven back to the Four Seasons by limousine. Four Seasons also paid for their rooms at the other hotel. It was the equivalent of a full money-back guarantee, and more.

Does this mean that every company that performs at the level of a Four Seasons need not offer a service guarantee? Could Federal Express, for example, drop its "absolutely, positively" assurance with little or no effect? Probably not. Its guarantee is such a part of its image that dropping the guarantee would hurt it.

In general, organizations that meet the following tests probably have little to gain by offering a service guarantee: the company is perceived by the market to be the quality leader in its industry; every employee is inculcated with the "absolute customer satisfaction" philosophy; employees are empowered to take whatever corrective action is necessary to handle complaints; errors are few; and a stated guarantee would be at odds with the company's image.

It is probably unnecessary to point out that few service companies meet these tests.

External Variables. Service guarantees may also be impractical where customer satisfaction is influenced strongly by external forces the service provider can't control. While everybody thinks their businesses are in this fix, most are wrong.

How many variables are truly beyond management's control? Not the work force. Not equipment

An airline can't guarantee on-time flights—but it *can* promise courtesy.

problems. Not vendor quality. And even businesses subject to "acts of God" (like weather) can control a great deal of their service quality.

BBBK is an example of how one company turned the situation around by analyzing the elements of the service-delivery process. By asking, "What obstacles stand in the way of our guaranteeing pest elimination?" Al Burger discovered that clients' poor cleaning and storage practices were one such obstacle. So the company requires customers to maintain sanitary practices and in some cases even make physical changes to their property (like putting in walls). By changing the process, the company could guarantee the outcome.

There may well be uncontrollable factors that create problems. As I noted earlier, such things as flight controllers, airport capacity, and weather limit the extent to which even the finest airline can consistently deliver on-time service. But how employees respond to such externally imposed problems strongly influences customer satisfaction, as British Airways executives learned from their market survey. When things go wrong, will employees go the extra yard to handle the problem? Why couldn't an airline that has refined its problem-handling skills to a science ensure absolute customer satisfaction – uncontrollable variables be damned? How many customers would invoke a guarantee if they understood that the reasons for a problem were completely out of the airline's control – if they were treated with warmth, compassion, and a sense of humor, and if the airline's staff communicated with them honestly?

Cheating. Fear of customer cheating is another big hurdle for most service managers considering offer-

> ## A guarantee can generate breakthrough service and change an industry.

ing guarantees. When asked why Lufthansa's guarantee required customers to present written proof of purchase, a manager at the airline's U.S. headquarters told me, "If we didn't ask for written proof, our customers would cheat us blind."

But experience teaches a different lesson. Sure, there will be cheats – the handful of customers who take advantage of a guarantee to get something for nothing. What they cost the company amounts to very little compared to the benefits derived from a strong guarantee. Says Michael Leven, a hotel industry executive, "Too often management spends its time worrying about the 1% of people who might cheat the company instead of the 99% who don't."

Phil Bressler of Domino's argues that customers cheat only when *they* feel cheated: "If we charge $8 for a pizza, our customers expect $8 worth of product and service. If we started giving them $7.50 worth of product and service, then they'd start looking for ways to get back that extra 50 cents. Companies create the incentive to cheat, in almost all cases, by cutting costs and not providing value."

Where the potential for false claims is high, a no-questions-asked guarantee may appear to be fool-hardy. When Domino's first offered its "delivery within 30 minutes or the pizza is free" guarantee, some college students telephoned orders from hard-to-find locations. The result was free pizza for the students and lost revenue for Domino's. In this environment, the guarantee was problematic because some students perceived it as a game against Domino's. But Bressler takes the view that the revenue thus lost was an investment in the future. "They'll be Domino's customers for life, those kids," he says.

High Costs. Managers are likely to worry about the costs of a service-guarantee program, but for the wrong reasons. Quality "guru" Philip Crosby coined the phrase "quality is free" (in his 1979 book, *Quality Is Free*) to indicate *not* that quality-improvement efforts cost nothing but that the benefits of quality improvement – fewer errors, higher productivity, more repeat business – outweigh the costs over the long term.

Clearly, a company whose operations are slipshod (or out of control) should not consider offering an unconditional guarantee; the outcome would be either bankruptcy from staggering payouts or an employee revolt stemming from demands to meet standards that are beyond the organization's capability. If your company is like most, however, it's not in that shape; you will probably only need to buttress operations somewhat. To be sure, an investment of financial and human resources to shore up weak points in the delivery system will likely cause a quick, sharp rise in expenditures.

How sharp an increase depends on several factors: your company's weaknesses (how far does it have to go to become good?), the nature of the industry, and the strength of your competition, for example. A small restaurant might simply spend more on employee recruiting and training, and perhaps on sponsoring quality circles; a large utility company might need to restructure its entire organization to overcome years of bad habits if it is to deliver on a guarantee.

Even though a guarantee carries costs, bear in mind that, as Crosby asserts, a badly performed service also incurs costs – failure costs, which come in many forms, including lost business from disgruntled consumers. In a guarantee program, you shift from spending to mop up failures to spending on preventing failures. And many of those costs are incurred in most organizations anyway (like outlays for staff time spent in planning meetings). It's just that they're spent more productively.

Breakthrough Service

One great potential of a service guarantee is its ability to change an industry's rules of the game by changing the service-delivery process as competitors conceive it.

BBBK and Federal Express both redefined the meaning of service in their industries, performing at levels that other companies have so far been unable to match. (According to the owner of a competing pest-control company, BBBK "is number one. There is no number two.") By offering breakthrough service, these companies altered the basis of competition in their businesses and put their competitors at a severe disadvantage.

What are the possibilities for replicating their success in other service businesses? Skeptics might claim that BBBK's and Federal Express's success is not widely applicable because they target price-insensitive customers willing to pay for superior service – in short, that these companies are pursuing differentiation strategies.

It is true that BBBK's complex preparation, cleaning, and checkup procedures are much more time consuming than those of typical pest-control operators, that the company spends more on pesticides than competitors do, and that its employees are well compensated. And many restaurants and hotels are willing to pay BBBK's higher prices because to them it's ultimately cheaper: the cost of "errors" (guests' spotting roaches or ants) is higher than the cost of error prevention.

But, because of the "quality is free" dictum, breakthrough service does not mean you must become the high-cost producer. Manpower's procedures are not radically more expensive than its competitors';

they're simply better. The company's skills-testing methods and customer-needs diagnoses surely cost less in the long run than a sloppy system. A company that inadequately screens and trains temporary-worker recruits, establishes no detailed customer specifications, and fails to check worker performance loses customers.

Manpower spends heavily on ways to reduce errors further, seeing this spending as an investment that will (a) protect its market position; (b) reduce time-consuming service errors; and (c) reinforce the company's values to employees. Here is the "absolute customer satisfaction" philosophy at work, and whatever cost increase Manpower incurs it makes up in sales volume.

Organizations that figure out how to offer – and deliver – guaranteed, breakthrough service will have tapped into a powerful source of competitive advantage. Doing so is no mean feat, of course, which is precisely why the opportunity to build a competitive advantage exists. Though the task is difficult, it is clearly not impossible, and the service guarantee can play a fundamental role in the process. ▽

Author's note: I thank Dan Maher for assistance in researching and writing this article.

Reprint 88405

"Actually, I don't want to make a deposit or a withdrawal
I just wanted to make sure everything was, you know, fine."

To learn how to keep customers,
track the ones you lose.

Zero Defections: Quality Comes to Services

by Frederick F. Reichheld and W. Earl Sasser, Jr.

The *real* quality revolution is just now coming to services. In recent years, despite their good intentions, few service company executives have been able to follow through on their commitment to satisfy customers. But service companies are beginning to understand what their manufacturing counterparts learned in the 1980s – that quality doesn't improve unless you measure it. When manufacturers began to unravel the costs and implications of scrap heaps, rework, and jammed machinery, they realized that "quality" was not just an invigorating slogan but the most profitable way to run a business. They made "zero defects" their guiding light, and the quality movement took off.

Service companies have their own kind of scrap heap: customers who will not come back. That scrap heap too has a cost. As service businesses start to measure it, they will see the urgent need to reduce it. They will strive for "zero defections" – keeping every customer the company can profitably serve – and they will mobilize the organization to achieve it.

Customer defections have a surprisingly powerful impact on the bottom line. They can have more to do with a service company's profits than scale, market share, unit costs, and many other factors usually associated with competitive advantage. As a customer's relationship with the company lengthens, profits rise. And not just a little. Companies can boost profits by almost 100% by retaining just 5% more of their customers.

Defecting customers send a clear signal: profit slump ahead.

While defection rates are an accurate leading indicator of profit swings, they do more than passively indicate where profits are headed. They also direct managers' attention to the specific things that are causing customers to leave. Since companies do not hold customers captive, the only way they can pre-

Frederick F. Reichheld is director of Bain & Company and leader of the firm's customer-retention practice. W. Earl Sasser, Jr. is a professor at the Harvard Business School.

vent defections is to outperform the competition continually. By soliciting feedback from defecting customers, companies can ferret out the weaknesses that really matter and strengthen them before profits start to dwindle. Defection analysis is therefore a guide that helps companies manage continuous improvement.

Charles Cawley, president of MBNA America, a Delaware-based credit card company, knows well how customer defections can focus a company's attention on exactly the things customers value. One morning in 1982, frustrated by letters from unhappy customers, he assembled all 300 MBNA employees and announced his determination that the company satisfy and keep each and every customer. The company started gathering feedback from defecting customers. And it acted on the information, adjusting products and processes regularly.

As quality improved, fewer customers had reason to leave. Eight years later, MBNA's defection rate is one of the lowest in its industry. Some 5% of its customers leave each year—half the average rate for the rest of the industry. That may seem like a small difference, but it translates into huge earnings. Without making any acquisitions, MBNA's industry ranking went from 38 to 4, and profits have increased sixteenfold.

The Cost of Losing a Customer

If companies knew how much it really costs to lose a customer, they would be able to make accurate evaluations of investments designed to retain customers. Unfortunately, today's accounting systems do not capture the value of a loyal customer. Most systems focus on current period costs and revenues and ignore expected cash flows over a customer's lifetime. Served correctly, customers generate increasingly more profits each year they stay with a company. Across a wide range of businesses, the pattern is the same: the longer a company keeps a customer, the more money it stands to make. (See the bar charts depicting "How Much Profit a Customer Generates over Time.") For one auto-service company, the expected profit from a fourth-year customer is more than triple the profit that same customer generates in the first year. When customers defect, they take all that profit-making potential with them.

It may be obvious that acquiring a new customer entails certain one-time costs for advertising, promotions, and the like. In credit cards, for example, companies spend an average of $51 to recruit a customer and set up the new account. But there are many more pieces to the profitability puzzle.

To continue with the credit card example, the newly acquired customers use the card slowly at first and generate a base profit. But if the customers stay a second year, the economics greatly improve. As they become accustomed to using the credit card and are satisfied with the service it provides, customers use it more and balances grow. In the second year—and the years thereafter—they purchase even more, which turns profits up sharply. We found this trend in each of the more than 100 companies in two dozen industries we have analyzed. For one industrial distributor, net sales per account continue to rise into the nineteenth year of the relationship.

As purchases rise, operating costs decline. Checking customers' credit histories and adding them to the corporate database is expensive, but those things need be done only once. Also, as the company gains experience with its customers, it can serve them more efficiently. One small financial consulting business that depends on personal relationships with cli-

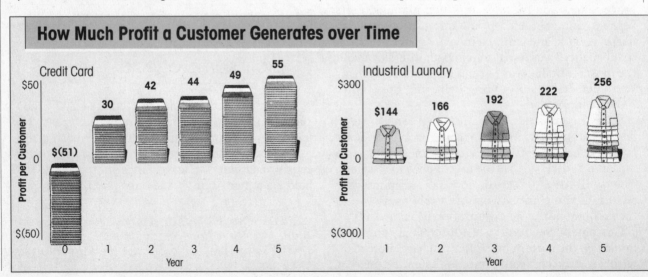

How Much Profit a Customer Generates over Time

Credit Card

$(51), 30, 42, 44, 49, 55 — Profit per Customer / Year (0, 1, 2, 3, 4, 5)

Industrial Laundry

$144, 166, 192, 222, 256 — Profit per Customer / Year (1, 2, 3, 4, 5)

ents has found that costs drop by two-thirds from the first year to the second because customers know what to expect from the consultant and have fewer questions or problems. In addition, the consultants are more efficient because they are familiar with the customer's financial situation and investment preferences.

Also, companies with long-time customers can often charge more for their products or services. Many people will pay more to stay in a hotel they know or to go to a doctor they trust than to take a chance on a less expensive competitor. The company that has developed such a loyal following can charge a premium for the customer's confidence in the business.

Yet another economic boon from long-time customers is the free advertising they provide. Loyal customers do a lot of talking over the years and drum up a lot of business. One of the leading home builders in the United States, for example, has found that more than 60% of its sales are the result of referrals.

These cost savings and additional revenues combine to produce a steadily increasing stream of profits over the course of the customer's relationship with the company. (See the chart "Why Customers Are More Profitable over Time.") While the relative importance of these effects varies from industry to industry, the end result is that longer term customers generate increasing profits.

To calculate a customer's real worth, a company must take all of these projected profit streams into account. If, for instance, the credit card customer leaves after the first year, the company takes a $21 loss. If the company can keep the customer for four more years, his or her value to the company rises sharply. It is equal to the net present value of the profit streams in the first five years, or about $100.

When a company lowers its defection rate, the average customer relationship lasts longer and profits climb steeply. One way to appreciate just how responsive profits are to changes in defection rates is to draw a defection curve. (See the graph, "A Credit Card Company's Defection Curve.") This shows clearly how small movements in a company's defection rate can produce very large swings in profits.

The curve shows, for example, that as the credit card company cuts its defection rate from 20% to 10%, the average life span of its relationship with a

> **Customers who defect to the competition can tell you exactly what parts of the business you must improve.**

customer doubles from five years to ten and the value of that customer more than doubles—jumping from $134 to $300. As the defection rate drops another 5%, the average life span of a customer relationship doubles again and profits rise 75%—from $300 to $525.

The credit card business is not unique. Although the shape of defection curves vary across industries, in general, profits rise as defection rates fall. Reducing defections by just 5% generated 85% more profits in one bank's branch system, 50% more in an insurance brokerage, and 30% more in an auto-service chain. (See the chart "Reducing Defections 5% Boosts Profits 25% to 85%.") MBNA America has found that a 5% improvement in defection rates increases its average customer value by more than 125%.

Understanding the economics of defections is useful to managers in several ways. For one thing, it shows that continuous improvement in service quality is not a cost but an investment in a customer who generates more profit than the margin on a one-time sale. Executives can therefore justify giving priority

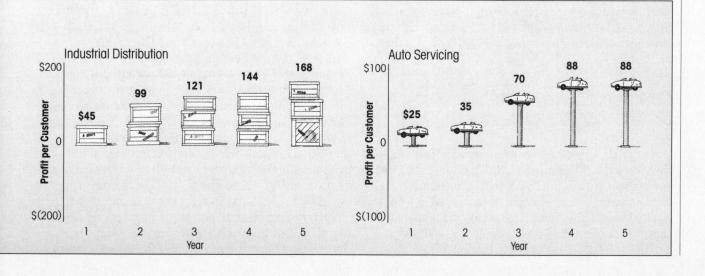

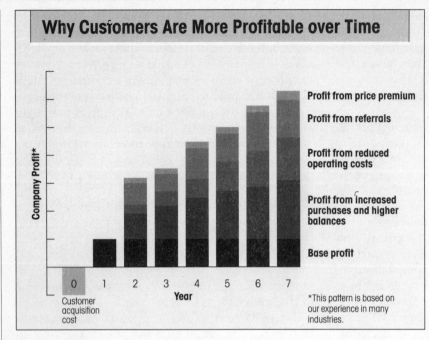

Why Customers Are More Profitable over Time

Company Profit*

Profit from price premium

Profit from referrals

Profit from reduced operating costs

Profit from increased purchases and higher balances

Base profit

0   1   2   3   4   5   6   7

Year

0 — Customer acquisition cost

*This pattern is based on our experience in many industries.

to investments in service quality versus things like cost reduction, for which the objectives have been more tangible.

Knowing that defections are closely linked to profits also helps explain why some companies that have relatively high unit costs can still be quite profitable. Companies with loyal, long-time customers can financially outperform competitors with lower unit costs and high market share but high customer churn. For instance, in the credit card business, a 10% reduction in unit costs is financially equivalent to a 2% decrease in defection rate. Low-defection strategies can overwhelm low-cost strategies.

And understanding the link between defections and profits provides a guide to lucrative growth. It is common for a business to lose 15% to 20% of its customers each year. Simply cutting defections in half will more than double the average company's growth rate. Companies with high retention rates that want to expand through acquisition can create value by acquiring low retention competitors and reducing their defections.

Defections Management

Although service companies probably can't—and shouldn't try to—eliminate all defections, they can and must reduce them. But even to approach zero defections, companies must pursue that goal in a coordinated way. The organization should be prepared to spot customers who leave and then to analyze and act on the information they provide.

Watch the door. Managing for zero defections requires mechanisms to find customers who have ended their relationship with the company—or are about to end it. While compiling this kind of customer data almost always involves the use of information technology of some kind, major investments in new systems are unnecessary.

The more critical issue is whether the business regularly gathers information about customers. Some companies already do. Credit card companies, magazine publishers, direct mailers, life insurers, cellular phone companies, and banks, for example, all collect reams of data as a matter of course. They have at their disposal the names and addresses, purchasing histories, and telephone numbers of all their customers. For these businesses, exposing defections is relatively easy. It's just a matter of organizing the data.

Sometimes, defining a "defection" takes some work. In the railroad business, for instance, few customers stop using your service completely, but a customer that shifts 80% of its shipments to trucks should not be considered "retained." The key is to identify the customer behaviors that both drive your economics and gauge customer loyalty.

For some businesses, the task of spotting defectors is challenging even if they are well defined, because customers tend to be faceless and nameless to management. Businesses like retailing will have to find creative ways to "know" their customers. Consider the example of Staples, the Boston-based office products discounter. It has done a superb job of gathering information usually lost at the cashier or sales clerk. From its opening, it had a database to store and analyze customer information. Whenever a customer goes through the checkout line, the cashier offers him or her a membership card. The card entitles the holder to special promotions and certain discounts. The only requirement for the card is that the person fill out an application form, which asks for things like name, job title, and address. All subsequent purchases are automatically logged against the card number. This way, Staples can accumulate detailed information about buying habits, frequency of visits, average dollar value spent, and particular items purchased.

Even restaurants can collect data. A crab house in Maryland, for instance, started entering into its PC information from the reservation list. Managers can now find out how often particular customers return and can contact those who seem to be losing interest in the restaurant.

What are defectors telling you? One reason to find customers who are leaving is to try to win them back. MBNA America has a customer-defection "swat" team staffed by some of the company's best telemarketers. When customers cancel their credit cards, the swat team tries to convince them to stay. It is successful half of the time.

But the more important motive for finding defectors is for the insight they provide. Customers who leave can provide a view of the business that is unavailable to those on the inside. And whatever caused one individual to defect may cause many others to follow. The idea is to use defections as an early warning signal – to learn from defectors why they left the company and to use that information to improve the business.

Unlike conventional market research, feedback from defecting customers tends to be concrete and specific. It doesn't attempt to measure things like attitudes or satisfaction, which are changeable and subjective, and it doesn't raise hypothetical questions, which may be irrelevant to the respondents. Defections analysis involves specific, relevant questions about why a customer has defected. Customers are usually able to articulate their reasons, and some skillful probing can get at the root cause.

This information is useful in a variety of ways, as the Staples example shows. Staples constantly tracks defections, so when customers stop doing business there or don't buy certain products, the store notices it immediately and calls to get feedback. It may be a clue that the competition is underpricing Staples on certain goods – a competitive factor management can explore further. If it finds sufficient evidence, Staples may cut prices on those items. This information is highly valued because it pinpoints the uncompetitive products and saves the chain from launching expensive broad-brush promotions pitching everything to everybody.

Staples's telemarketers try to discern which merchandise its customers want and don't want and why. The company uses that information to change its buying stock and to target its catalogs and coupons more precisely. Instead of running coupons in the newspaper, for instance, it can insert them in the catalogs it sends to particular customers or industries that have proved responsive to coupons.

Defections analysis can also help companies decide which service-quality investments will be profitable. Should you invest in computerized cash registers or a new phone system? Which of the two will address the most frequent causes of defection? One bank made a large investment to improve the accuracy of monthly account statements. But when the bank began to study defectors, it learned that less than 1% of its customers were leaving because of inaccurate statements.

A company that is losing customers because of long lines can estimate what percentage of defectors it would save by buying new cash registers, and it can use its defection curve to find the dollar value of saving them. Then, using standard investment-analysis techniques, it can compare the cost of the new equipment with the benefit of keeping customers.

Achieving service quality doesn't mean slavishly keeping all customers at any cost. There are some customers the company should not try to serve. If particular types of customers don't stay and become profitable, companies should not invest in attracting them. When a health insurance company realized that certain companies purchase only on the basis of price and switch health insurers every year, for example, it decided not to waste its efforts seeking their business. It told its brokers not to write policies for companies that have switched carriers more than twice in the past five years.

Conversely, much of the information used to find defectors can point to common traits among custom-

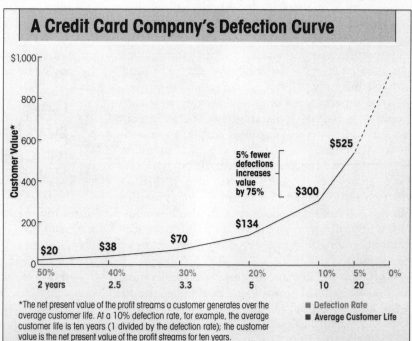

A Credit Card Company's Defection Curve

*The net present value of the profit streams a customer generates over the average customer life. At a 10% defection rate, for example, the average customer life is ten years (1 divided by the defection rate); the customer value is the net present value of the profit streams for ten years.

- Defection Rate
- Average Customer Life

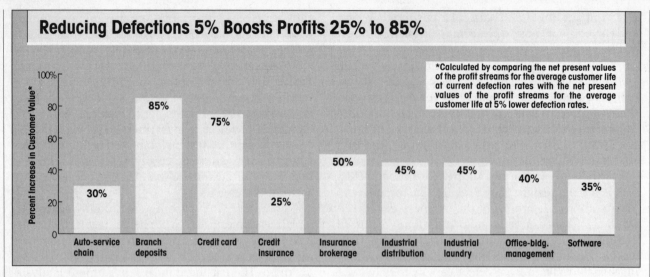

Reducing Defections 5% Boosts Profits 25% to 85%

*Calculated by comparing the net present values of the profit streams for the average customer life at current defection rates with the net present values of the profit streams for the average customer life at 5% lower defection rates.

Percent Increase in Customer Value*

Auto-service chain	Branch deposits	Credit card	Credit insurance	Insurance brokerage	Industrial distribution	Industrial laundry	Office-bldg. management	Software
30%	85%	75%	25%	50%	45%	45%	40%	35%

ers who stay longer. The company can use defection rates to clarify the characteristics of the market it wants to pursue and target its advertising and promotions accordingly.

The Zero Defections Culture

Many business leaders have been frustrated by their inability to follow through on their public commitment to service quality. Since defection rates are measurable, they are manageable. Managers can establish meaningful targets and monitor progress. But like any important change, managing for zero defections must have supporters at all organizational levels. Management must develop that support by training the work force and using defections as a primary performance measure.

Everyone in the organization must understand that zero defections is the goal. Mastercare, the auto-service subsidiary of Bridgestone/Firestone, emphasizes the importance of keeping customers by stating it clearly in its mission statement. The statement says, in part, that the company's goal is "to provide the service-buying public with a superior buying experience that will encourage them to return willingly and to share their experience with others." MBNA America sends its paychecks in envelopes labeled "Brought to you by the customer." It also has a customer advocate who sits in on all major decision-making sessions to make sure customers' interests are represented.

It is important to make all employees understand the lifetime value of a customer. Phil Bressler, the co-owner of five Domino's Pizza stores in Montgomery County, Maryland, calculated that regular customers were worth more than $5,000 over the life of a ten-

year franchise contract. He made sure that every order taker, delivery person, and store manager knew that number. For him, telling workers that customers were valuable was not nearly as potent as stating the dollar amount: "It's so much more than they think that it really hits home."

Mastercare has redesigned its employee training to emphasize the importance of keeping customers. For example, many customers who stopped doing business with Mastercare mentioned that they didn't like being pressured into repairs they had not planned on. So Mastercare now trains store managers to identify and solve the customer's problem rather than to maximize sales. Videos and role-playing dramatize these different definitions of good service.

Mastercare's message to employees includes a candid admission that previous, well-intentioned incentives had inadvertently caused employees to run the business the wrong way; now it is asking them to change. And it builds credibility among employees by sharing its strategic goals and customer outreach

> Great-West Life Assurance pays brokers a premium for lowering customer defection rates.

plans. In the two target markets where this approach has been used, results are good. Employees have responded enthusiastically, and 25% more customers say they intend to return.

Senior executives at MBNA America learn from defecting customers. Each one spends four hours a month in a special "listening room" monitoring routine customer service calls as well as calls from customers who are canceling their credit cards.

Beyond conveying a sense of urgency, training should teach employees the specifics of defections analysis, like how to gather the information, whom to pass it on to, and what actions to take in response. In one company's branch banking system, retention data is sent monthly to the regional vice presidents and branch managers for review. It allows the regional vice presidents to identify and focus on branches that most need to improve service quality, and it gives branch managers quick feedback on performance.

Employees will be more motivated if incentives are tied to defection rates. MBNA, for example, has determined for each department the one or two things that have the biggest impact on keeping customers. Each department is measured daily on how well performance targets are met. Every morning, the previous day's performance is posted in several places throughout the building. Each day that the company hits 95% of these performance targets, MBNA contributes money to a bonus pool. Managers use the pool to pay yearly bonuses of up to 20% of a person's salary. The president visits departments that fall short of their targets to find out where the problem lies.

Great-West Life Assurance Company of Englewood, Colorado also uses incentives effectively. It pays a 50% premium to group-health-insurance brokers that hit customer-retention targets. This system gives brokers the incentive to look for customers who will stay with the company for a long time.

Having everyone in the company work toward keeping customers and basing rewards on how well they do creates a positive company atmosphere. Encouraging employees to solve customer problems and eliminate the source of complaints allows them to be "nice," and customers treat them better in return. The overall exchange is more rewarding, and people enjoy their work more. Not just customers but also employees will want to continue their relationship with the business. MBNA is besieged by applicants for job openings, while a competitor a few miles away is moving some of its operations out of the state because it can't find enough employees.

The success of MBNA shows that it is possible to achieve big improvements in both service quality and profits in a reasonably short time. But it also shows that focusing on keeping customers instead of simply having lots of them takes effort. A company

> **Employees like to work for a company that keeps its customers. MBNA America has far more applicants than jobs.**

can leverage business performance and profits through customer defections only when the notion permeates corporate life and when all organizational levels understand the concept of zero defections and know how to act on it.

Trying to retain all of your profitable customers is elementary. Managing toward zero defections is revolutionary. It requires careful definition of defection, information systems that can measure results over time in comparison with competitors, and a clear understanding of the microeconomics of defection.

Ultimately, defections should be a key performance measure for senior management and a fundamental component of incentive systems. Managers should know the company's defection rate, what happens to profits when the rate moves up or down, and why defections occur. They should make sure the entire organization understands the importance of keeping customers and encourage employees to pursue zero defections by tying incentives, planning, and budgeting to defection targets. Most important, managers should use defections as a vehicle for continuously improving the quality and value of the services they provide to customers.

Just as the quality revolution in manufacturing had a profound impact on the competitiveness of companies, the quality revolution in services will create a new set of winners and losers. The winners will be those who lead the way in managing toward zero defections.   ▽

Reprint 90508

GROWING CONCERNS

The key to growth and profits is getting employees to give away more of your products and services.

My Employees Are My Service Guarantee

by Timothy W. Firnstahl

I own a chain of four restaurants in and around Seattle, and my company exists for one reason only – to make other people happy. Every time a customer leaves one of our restaurants with a more optimistic view of the world, we've done our job. Every time we fail to raise a customer's spirits with good food, gratifying service, and a soothing atmosphere, we haven't done our job.

To the extent that we satisfy customers, we fulfill our company goal. This observation may seem self-evident and trivial – a useful motto, a business axiom that a lot of business-people understandably overlook in the day-to-day flood of details – but I have found it the very key to growth and profits. And after much trial and error, I have come up with a strategy for ensuring customer satisfaction that has worked wonders in our business and can, I'm convinced, work wonders in other businesses as well.

It starts with a guarantee – not that moth-eaten old promise of a cheerful refund – but a guarantee that customers will be satisfied with their whole experience of the company's products and services. It moves on to a system for giving employees complete responsibility and *authority* for making the guarantee stick. It ends with a process for identifying system failures – the problems in organization, training, and other internal programs that cause customer dissatisfaction.

I call the whole thing "ultimate strategy." That may sound pretentious. But because it redefines a company's ultimate reason for being and succeeding, and because it underlines the importance of finding the ultimate causes of every system failure, I think the name is justified.

Service with a smile and a seed of doubt

Ultimate strategy had its origins in the success of a restaurant business I cofounded ten years ago. (I recently started a different restaurant

business, but the strategy hasn't changed.) The first restaurant, specializing in steaks and featuring a huge bar, went over so well that we opened another. Five years ago, we had three restaurants, $7.5 million in sales, and moderate profits. Clearly, many of our customers were satisfied.

But I was bothered by what I saw as an unacceptable level of complaints and by our haphazard responses to them. Not that we didn't try. We happily apologized and gave a free dessert to any customer who complained about slow service, and we cheerfully picked up the cleaning bill when one of our employees spilled the soup. Customers who wrote in to complain about reservations mix-ups or rude service got certificates for complimentary meals.

It was just that our procedures for responding seemed all wrong. Giving out that free dessert required approval from a manager. Getting a suit cleaned meant filling out a form and getting a manager to sign it. I also didn't like the idea that people had to write us with their complaints before we made amends. And I wasn't convinced that a free meal was enough.

Moreover, our response to complaints didn't appear to have any effect on the number or type of complaints we received, most of which concerned speed of service and quality of food. And it wasn't the employees' fault. They knew complaints had top priority, but they didn't know how to respond to them. We were all on a treadmill, getting nowhere.

The guarantee

Then five years ago, when the book *In Search of Excellence* was in vogue, I spent considerable time writing Ten Tenets of Excellence for our organization. We included them in our train-

Founder and CEO of Satisfaction Guaranteed Eateries, Inc. in Seattle, Timothy W. Firnstahl is a restaurant zealot and the fourth generation of his family in the food industry. Previously, he wrote for HBR on how entrepreneurs can delegate responsibility.

ing manuals and posted them in the restaurants and the offices. One day about a year later, someone asked me what the sixth tenet was, and I couldn't tell her. It came to me that if I couldn't remember the Ten Tenets of Excellence, surely no one else could either. That meant the company had no strategy known to its employees.

So I hit on something simpler and more compelling—the guarantee. We expressed it as a promise: *Your Enjoyment Guaranteed. Always.* As a company rallying cry, it seemed to work much better than the Tenets of Excellence. Cryptic mission statements, unreviewed strategic plans, the hidden dreams of management: all that gave way to a company game plan—customer satisfaction—that everyone could understand and remember and act on. For the first time, employees and management had a strategy in common.

Your Enjoyment Guaranteed. Always. This promise became our driving force. We included it in all our advertising. We printed it on every menu, letterhead, and guest check. To make it live for our employees, we did a series of internal promotions. We reduced it to an acronym, YEGA, and posted it everywhere for employees to see.

What good is a guarantee that makes complaining an ordeal for the customer?

We held a series of meetings, where we found workers receptive to both the acronym and the simplicity of the idea. Each of our 600 employees signed a contract pledging YEGA follow-through. We created a YEGA logo and put it everywhere, on report forms, on training manuals, on wall signs. We started the *YEGA News* and distributed YEGA pins, shirts, name tags, even underwear. We announced that failure to enforce YEGA would be cause for dismissal.

For a year or so, YEGA dominated the company's consciousness. But as time went by, I grew increasingly uncomfortable. Complaints were coming in at the same old rate. I could see the guarantee being implemented here and there, now and then, but not on a regular, companywide basis. I'd run into another brick wall.

Empowering employees

One evening about two years back as I was driving home from work, the cause of the problem hit me. The guarantee by itself wasn't enough. We had given employees responsibility without giving them authority. The result was that they tried to bury mistakes or blame others. I saw it every time we tried to track down a complaint. The food servers blamed the kitchen for late meals. The kitchen blamed the food servers for placing orders incorrectly.

Problems inevitably crop up in a busy restaurant, and when a customer grumbles the tendency is to gloss over the complaint with pleasantness. Follow-through means fetching the manager or filling out forms or both. Climbing the ladder of hierarchical approvals is simply too frustrating and time-consuming —for customer as well as employee.

For our guarantee to be truly effective, we needed to give workers themselves the power to make good on the promise of the guarantee—at once and on the spot. Eliminate the hassle for the customer and for ourselves. No forms to fill out, no phone calls to make, no 40 questions to answer, just immediate redress by the closest employee.

So I instituted the idea that employees could and should do *anything* to keep the customer happy. In the event of an error or delay, any employee right down to the busboy could provide complimentary wine or desserts, or pick up an entire tab if necessary.

Of course, we provided some guidelines. For instance, when guests have to wait more than 10 minutes beyond their reservation time, but less than 20, we suggest free drinks. If they wait more than 20 minutes, the entire meal might be free. If the bread arrives more than 5 minutes after

the guests sit down, we suggest free clam chowder. And so forth, using what we know to be optimum intervals for most orders.

At the same time, we urged employees not to get bogged down in the guidelines. The last thing we wanted was nitpicking: "OK, I got them the bread in five minutes exactly. Do I just apologize, or do they get clam chowder?" Satisfaction does not mean quibbling—it means a contented customer. Different guests respond in different ways, so we told our employees not to feel limited by the guidelines and to do whatever it took to make sure guests enjoyed themselves.

Employees were initially wary of their new authority. Never having had complete control, they were naturally hesitant and skeptical. It was hard to convince them they wouldn't be penalized for giving away free food and drinks.

But once they got used to the idea, employees liked knowing that the company believed so strongly in its products and services that it wholeheartedly stood behind its work— and theirs. They liked working for a restaurant known for its unhesitating commitment to customer satisfaction. Preeminence in any field gives people feelings of self-worth they could never get from just making a buck. Their power as company representatives increased their pride in the business, and that, in turn, increased motivation.

Once our employees overcame their skepticism, they quickly grew creative and aggressive in their approach to the guarantee. In one case, a customer wanted a margarita made the way a competitor made its. So our bartender called the bartender at the other restaurant and, bartender-to-bartender, learned the special recipe. In another case, an elderly woman who had not been in our restaurant for years ordered breakfast, which we no longer served. The waiter and the chef sent someone to the market for bacon and eggs and served the breakfast she wanted.

If the guarantee is really working the way it's supposed to, customers become less inhibited about complaining. Too often, customers hold their peace but vote with their feet by taking their business to the competition. The promise of the guarantee's enforcement stimulates them to help us expose our own failures.

We even asked for their criticism. Once a month, using reservations lists and credit card charges, groups of employees called several hundred customers and asked them to rate their experience. Were the food and service lousy, OK, good, very good, or excellent? If they said "OK," that meant "lousy" to us, and they got a letter of apology, a certificate for a free meal, and a follow-up phone call.

Aside from the data we gathered, the phone calls were great promotion. Most people were amazed and delighted that we took the trouble to phone them, and many developed enormous loyalty to our restaurants.

System-failure costs

Customer complaints are company failures and require immediate correction. So far so good. But corrections cost money. Free drinks and meals add up quickly.

Yet, paradoxically, spending money is the goal. Every dollar paid out to offset customer dissatisfaction is a signal that the company must change in some decisive way. The guarantee brings out a true, hard-dollars picture of company failures and forces us to assume full responsibility for our output. The cost of keeping a company's promises is not just the price tag on the guarantee, it is the cost of system failure. The money was spent because the product did not perform, and when the product fails to perform, the system that produced it is at fault.

A somnolent business can be rudely awakened by the magnitude of its system-failure costs. We certainly were. Our previous guarantee expenses doubled. The problems had always been there, hidden. Only the huge cost of the new strategy revealed that they were gutting profits. Suddenly, we had a real incentive to fix the systems that weren't working, since the alternatives—sacrificing profits permanently or restricting the power to enforce the guarantee—were both unacceptable.

> **Every dollar you give away is a plus—it puts your finger on a problem you can fix.**

Notice that system-failure costs are not the same as employee-failure costs. System-failure costs measure the extent of the confusion in company structure, for which management alone is to blame. By welcoming every guarantee payoff—every system-failure expense—as an otherwise lost insight, you can make every problem pay a dividend. The trick is to reject Band-Aid solutions, to insist on finding the ultimate cause of each problem, and then to demand and expect decisive change. (Another way to sugarcoat the pill of system-failure costs is to think of the free food and drinks as a word-of-mouth advertising budget. No one forgets to mention a free meal to a friend or neighbor.)

Our search for the culprit in a string of complaints about slow food service in one restaurant led first to the kitchen and then to one cook. But pushing the search one step further revealed several unrealistically complex dishes that no one could have prepared swiftly.

In another case, our kitchens were turning out wrong orders at a rate

---

The Hassle Factor

Imagine you've bought a new pair of shoes at a downtown store. A week later, one sole starts to come off, so you take them back. You drive downtown through heavy traffic and spend 15 minutes finding a parking place. You explain the problem to the salesman, who says, "We stand behind our merchandise." He gives you a new pair of shoes.

Question One: Are you happy?

Answer: Well, no, you're not. Sure, you got a new pair of shoes, and the salesman was pleasant enough, but you had to take time out of your day and go to a lot of trouble to get what you should have gotten in the first place. In short, the whole transaction was a hassle, and neither the salesman nor the store did anything to make it up to you.

Question Two: What should they have done?

Answer: Replace plus one. Besides giving you a new pair of shoes, the salesman should have thrown in a pair of socks or stockings to repay you for your hassle. Instead of, "We'll replace inferior merchandise whenever a customer complains," the store's message should be, "We really regret your inconvenience and want to make you happy."

Like the shoe store, we stand behind our products and services. Unlike the shoe store, we'll do more than the customer demands to make it right. If a guest doesn't like her salad, don't charge her for it. But what about the Hassle Factor?

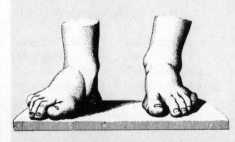

Replace plus one. By all means, give her the salad free of charge. But buy her a drink or dessert as well—or whatever else it takes to make her happy.

Adapted from the restaurant training manual

that was costing us thousands of dollars a month in wasted food. The cooks insisted that the food servers were punching incorrect orders into the kitchen printout computer. In times past, we might have ended our search right there, accused the food servers of sloppiness, and asked everyone to be more careful. But now, adhering to the principle of system failure not people failure, we looked beyond the symptoms and found a flaw in our training. We had simply never taught food servers to double-check their orders on the computer screen, and the system offered no reward for doing so. Mistakes plummeted as soon as we started training people properly and handing out awards each month for the fewest ordering errors and posting lists of the worst offenders (no punishments, just names).

Of course, correcting system failures is seldom an easy task. One way to avoid making problems worse is to audition problem solutions with small, quick-hit field tests. For exam-

What is a word-of-mouth advertising budget? It's the money you spend honoring your guarantee.

ple, we experimented with new service procedures at one station in one restaurant, or we offered new menu items as nonmenu specials, or we borrowed equipment for a test run before leasing or buying it. When we had a problem with coffee quality, we tried using expensive, high-quality vacuum carafes in one restaurant. Quality improved substantially (and waste was cut in half), so we adopted the thermoses in all our restaurants.

When some customers complained about our wine service, we realized that we gave the subject only

three pages in our employee manual. So we put together a training and motivation package that included instruction about the characteristics and distinctions of different wines, as well as a system of awards for selling them effectively. We also pointed out to our food servers that selling more wine increases the size of checks and thus of tips.

In short, honoring the guarantee has led to new training procedures, recipe and menu changes, restaurant redesign, equipment purchases, and whatever else it took to put things right and keep them right. In the long run, the guarantee works only if it reduces system-failure costs and increases customer satisfaction.[1]

This kind of problem solving is popular with employees. Since the object of change is always the company, employees don't get blamed for problems beyond their control.

As you find and correct the ultimate causes of your system failures, you can reasonably expect your profits to improve. But you can begin to tell if you're succeeding even before you see it on the bottom line. Remember, costs will go up before they come down, so high system-failure costs and low phone-survey complaint rates probably mean you're on the right track. Conversely, low system-failure costs and a high rate of "lousies" and "OKs" from customers almost certainly indicate that the promise is not being kept, that your expensive system failures are not getting corrected, and that your organization has yet to understand that customer satisfaction is the only reason for the company's existence.

Our own system-failure costs rose to a high of $40,000 a month two years ago and then fell to $10,000 a month. Meanwhile, sales rose 25%, profits doubled, and the cash in the bank grew two-and-a-half times.

Making it work

It is easier to give someone a bowl of clam chowder than a free CAT scanner or an industry marketing study, so of course the nature of the guarantee will change from business to business. Still, the point is not free food, the point is customer satisfac-

tion. It is always possible to satisfy the customer if the business is sufficiently committed to that goal.

Here are my suggestions for formulating your own ultimate strategy.

1. *Make the guarantee simple and easy to understand.* Think about the company's primary customer benefit

Firnstahl's first rule: Always deal with complaints before they're made.

and how you achieve it. In our case, the principal benefit is enjoyment. For many, it will be dependability. For others, cost or flexibility.

For the sake of impact, try to develop a guarantee that's memorable, maybe one that reduces to an acronym. The restaurants I now own use WAGS (We Always Guarantee Satisfaction), which I like even better than YEGA. Whatever you do, make it significant, simple, and unconditional. Think of these famous promises that changed whole companies:

"We try harder." (Avis)

"Absolutely Positively Overnight." (Federal Express)

Once you settle on a guarantee, commit to it for the long term. Continual change confuses the public and the organization. Plan to stick with a particular promise for at least five years.

2. *Make sure employees know how to use their new authority.* For most employees, full power and responsibility to put things right will be a new experience. After all, they're used to the old hierarchical approach. So it's up to you to make sure they don't underuse their power. In our training programs, we advise new employees to take action before the guest has to ask for a remedy. We don't want to make customers decide whether they're entitled to get something free—most people find that embarrassing. The food server should find the solution and present it to the guest as a done deal: "I'm sorry your drink wasn't prepared the

1. For more on guarantees, see Christopher W. L. Hart, "The Power of Unconditional Service Guarantees," HBR July-August 1988, p. 54.

way you like it. Of course, there will be no charge for that. And please accept these chowders on the house with my apologies."

We also insist that the customer is always right, even when the customer is wrong. Let's say a guest insists that all clam chowder has

> **Giving your employees total power and authority will scare them to death and make you rich.**

potatoes. He's wrong, but that's no excuse to make him look stupid. When we say, "The guest is never wrong," we mean a server should never question a guest's judgment and perception. Don't stand and argue about whether a steak is medium-rare or medium. Take it away and get one broiled the way the customer wants it.

The real issues are these: The guest is there to have a good time. The guest is in the employee's care.

Finally, we think that power and responsibility are not enough. Employees must also have rewards. Good thinking and positive action deserve money, praise, the limelight, advancement, and all the other encouragements a company can think of.

We spark employee thought and action by dividing a $10,000 bonus among the employees of each restaurant once its system-failure costs and phone-audit complaint rates drop to 25% of their all-time highs. Every month, we pay thousands of dollars in awards to employees who have helped to find and cure the ultimate causes of system failures. In effect, we commission everyone to change the organization for the better.

3. *Make progress visible.* Stay away from written progress reports— graphs communicate better. A creative in-house accountant can play with the data until it's readily under-

standable to everyone. We display our new WAGS graphics throughout the company for everyone to see.

In our experience, system-failure costs go through four phases.

Start: Employees are wary of using their new power and authority. Phone-audit complaint rates are high and system-failure costs are low.

Under way: Employees begin to believe in the organization's commitment to the guarantee. Phone-audit complaint rates are still high; system-failure costs start to rise.

Mid-point: Employees accept and act on the company promise. System-failure costs remain high. Phone-audit complaint rates start dropping as the company starts satisfying customers in earnest.

Success: The company has achieved elemental change and raised itself to a higher level of merit. System-failure costs and phone-audit complaint rates are both low.

In general, there is a roller coaster effect that tells you when ultimate strategy is working. Costs go up. Complaints go down. Sales go up. Costs go down. Profits go up.

One word of caution: you will never perfect your company's system. As long as you offer an absolute guarantee on your products and services, you will incur system-failure costs. There is always more work to do, and a CEO's personal commitment and persistence are often necessary to get it done. But motivated employees are essential.

People often ask us where we find such wonderful employees. While it's true that we screen carefully, I believe our employees are better than most because they have the power and the obligation to solve customer problems on their own and on the spot. Giving them complete discretion about how they do it has also given them pride. Many companies have tried so many different programs and gimmicks that employees have become cynical and indifferent. The people who work for us know we take our guarantee seriously—and expect them to do the same. We use the same ultimate strategy to satisfy both customers and employees. ▽

Reprint 89407

READ THE FINE PRINT

REPRINTS
Telephone: 617-495-6192
Fax: 617-495-6985

Current and past articles are available, as is an annually updated index. Discounts apply to large-quantity purchases.

Please send orders to
HBR Reprints
Harvard Business School
Publishing Division
Boston, MA 02163.

HOW CAN *HARVARD BUSINESS REVIEW* ARTICLES WORK FOR YOU?

For years, we've printed a microscopically small notice on the editorial credits page of the *Harvard Business Review* alerting our readers to the availability of *HBR* articles.

Now we invite you to take a closer look at some of the many ways you can put this hard-working business tool to work for you.

IN THE CORPORATE CLASSROOM.

There's no more effective, or cost-effective, way to supplement your corporate training programs than in-depth, incisive *HBR* articles.

Affordable and accessible, it's no wonder hundreds of companies and consulting organizations use *HBR* articles as a centerpiece for management training.

IN-BOX INNOVATION.

Where do your company's movers and shakers get their big ideas? Many find the inspiration for innovation in the pages of *HBR*. They then share the wealth and spread the word by distributing *HBR* articles to company colleagues.

IN MARKETING AND SALES SUPPORT.

HBR articles are a substantive leave-behind to your sales calls. And they can add credibility to your direct mail campaigns. They demonstrate that your company is on the leading edge of business thinking.

CREATE CUSTOM ARTICLES.

If you want to pack even greater power in your punch, personalize *HBR* articles with your company's name or logo. And get the added benefit of putting your organization's name before your customers.

AND THERE ARE 500 MORE REASONS IN THE *HBR CATALOG.*

In all, the *Harvard Business Review Catalog* lists articles on over 500 different subjects. Plus, you'll find books and videos on subjects you need to know.

The catalog is yours for just $8.00. To order *HBR* articles or the *HBR Catalog* (No. 21019), call 617-495-6192. Please mention telephone order code 025A when placing your order. Or FAX us at 617-495-6985.

And start putting *HBR* articles to work for you.

**Harvard Business School
Publications**

Call 617-495-6192 to order the *HBR Catalog.*

(Prices and terms subject to change.)